计算机科学与技术专业核心教材体系建设 —— 建议使用时间

课程系列	基础系列	电类系列	程序系列	系统系列	应用系列	选修系列
一年级上	大学计算机基础		计算机程序设计			
一年级下	信息安全导论	电子技术基础	面向对象程序设计 程序设计实践	计算机原理		
二年级上	离散数学(上)	数字逻辑设计 数字逻辑设计实验	数据结构	操作系统		
二年级下	离散数学(下)		算法设计与分析	计算机系统综合实践		
三年级上			编译原理	计算机网络	人工智能导论 数据库原理与技术 嵌入式系统	
三年级下			软件工程	计算机体系结构	计算机图形学	
四年级上			软件工程综合实践			机器学习 物联网导论 大数据分析技术 数字图像技术
四年级下						

面向新工科专业建设计算机系列教材

网络工程实验教程
微课版

杨恩宁 李联宁 胡 朋 许大炜 编著

清华大学出版社
北京

内 容 简 介

本书是《网络工程》(第3版)的配套实验教材,编者结合高校学生专业培养方案以及社会对人才的需求,依托十多年的教学经验,共设计了25个实验,分为基础篇和技能提升篇两篇。

基础篇共有15个实验,分别是认识网络、网线的制作、常用网络命令的使用、快速以太网的构建、网络设备IOS在CLI模式下的基本使用、交换机的常用配置与管理、基于交换机端口的VLAN划分、VLAN间路由设计、CDP的配置与验证、静态路由的配置与验证、默认路由的配置与验证、RIP的配置与验证、网络应用服务平台的构建与测试、小型无线网络的设计与配置、网络打印服务的配置。

技能提升篇共有10个实验。主要从交换与路由的高级应用出发,结合网络安全应用以及典型企业网络的设计与实现几方面考虑,进行实验的设计和实现。

本书实用性强,应用范围广,适合计算机类和电子信息类专业学生使用,可作为高等院校本科教材,对于计算机类工程技术人员也有学习和参考价值。

本书封面贴有清华大学出版社防伪标签,无标签者不得销售。
版权所有,侵权必究。举报:010-62782989,beiqinquan@tup.tsinghua.edu.cn。

图书在版编目(CIP)数据

网络工程实验教程:微课版/杨恩宁等编著. —北京:清华大学出版社,2023.8
面向新工科专业建设计算机系列教材
ISBN 978-7-302-64059-2

Ⅰ.①网… Ⅱ.①杨… Ⅲ.①网络工程-实验-高等学校-教材 Ⅳ.①TP393-33

中国国家版本馆CIP数据核字(2023)第126935号

责任编辑:白立军
封面设计:刘 键
责任校对:郝美丽
责任印制:曹婉颖

出版发行:清华大学出版社
    网　　址:http://www.tup.com.cn,http://www.wqbook.com
    地　　址:北京清华大学学研大厦A座　　邮　编:100084
    社 总 机:010-83470000　　邮　购:010-62786544
    投稿与读者服务:010-62776969,c-service@tup.tsinghua.edu.cn
    质量反馈:010-62772015,zhiliang@tup.tsinghua.edu.cn
    课件下载:http://www.tup.com.cn,010-83470236
印 装 者:三河市龙大印装有限公司
经　　销:全国新华书店
开　　本:185mm×260mm　　印 张:13　　插 页:1　　字 数:318千字
版　　次:2023年8月第1版　　印　次:2023年8月第1次印刷
定　　价:49.00元

产品编号:097203-01

出版说明

一、系列教材背景

人类已经进入智能时代，云计算、大数据、物联网、人工智能、机器人、量子计算等是这个时代最重要的技术热点。为了适应和满足时代发展对人才培养的需要，2017年2月以来，教育部积极推进新工科建设，先后形成了"复旦共识""天大行动"和"北京指南"，并发布了《教育部高等教育司关于开展新工科研究与实践的通知》《教育部办公厅关于推荐新工科研究与实践项目的通知》，全力探索形成领跑全球工程教育的中国模式、中国经验，助力高等教育强国建设。新工科有两个内涵：一是新的工科专业；二是传统工科专业的新需求。新工科建设将促进一批新专业的发展，这批新专业有的是依托于现有计算机类专业派生、扩展而成的，有的是多个专业有机整合而成的。由计算机类专业派生、扩展形成的新工科专业有计算机科学与技术、软件工程、网络工程、物联网工程、信息管理与信息系统、数据科学与大数据技术等。由计算机类学科交叉融合形成的新工科专业有网络空间安全、人工智能、机器人工程、数字媒体技术、智能科学与技术等。

在新工科建设的"九个一批"中，明确提出"建设一批体现产业和技术最新发展的新课程""建设一批产业急需的新兴工科专业"。新课程和新专业的持续建设，都需要以适应新工科教育的教材作为支撑。由于各个专业之间的课程相互交叉，但是又不能相互包含，所以在选题方向上，既考虑由计算机类专业派生、扩展形成的新工科专业的选题，又考虑由计算机类专业交叉融合形成的新工科专业的选题，特别是网络空间安全专业、智能科学与技术专业的选题。基于此，清华大学出版社计划出版"面向新工科专业建设计算机系列教材"。

二、教材定位

教材使用对象为"211工程"高校或同等水平及以上高校计算机类专业及相关专业学生。

三、教材编写原则

(1) 借鉴 *Computer Science Curricula* 2013(以下简称 CS2013)。CS2013

的核心知识领域包括算法与复杂度、体系结构与组织、计算科学、离散结构、图形学与可视化、人机交互、信息保障与安全、信息管理、智能系统、网络与通信、操作系统、基于平台的开发、并行与分布式计算、程序设计语言、软件开发基础、软件工程、系统基础、社会问题与专业实践等内容。

(2) 处理好理论与技能培养的关系,注重理论与实践相结合,加强对学生思维方式的训练和计算思维的培养。计算机专业学生能力的培养特别强调理论学习、计算思维培养和实践训练。本系列教材以"重视理论,加强计算思维培养,突出案例和实践应用"为主要目标。

(3) 为便于教学,在纸质教材的基础上,融合多种形式的教学辅助材料。每本教材可以有主教材、教师用书、习题解答、实验指导等。特别是在数字资源建设方面,可以结合当前出版融合的趋势,做好立体化教材建设,可考虑加上微课、微视频、二维码、MOOC等扩展资源。

四、教材特点

1. 满足新工科专业建设的需要

系列教材涵盖计算机科学与技术、软件工程、物联网工程、数据科学与大数据技术、网络空间安全、人工智能等专业的课程。

2. 案例体现传统工科专业的新需求

编写时,以案例驱动,任务引导,特别是有一些新应用场景的案例。

3. 循序渐进,内容全面

讲解基础知识和实用案例时,由简单到复杂,循序渐进,系统讲解。

4. 资源丰富,立体化建设

除了教学课件外,还可以提供教学大纲、教学计划、微视频等扩展资源,以方便教学。

五、优先出版

1. 精品课程配套教材

主要包括国家级或省级的精品课程和精品资源共享课的配套教材。

2. 传统优秀改版教材

对于已经出版、得到市场认可的优秀教材,由于新技术的发展,计划给图书配上新的教学形式、教学资源的改版教材。

3. 前沿技术与热点教材

反映计算机前沿和当前热点的相关教材,例如云计算、大数据、人工智能、物联网、网络空间安全等方面的教材。

六、联系方式

联系人：白立军
联系电话：010-83470179
联系和投稿邮箱：bailj@tup.tsinghua.edu.cn

面向新工科专业建设计算机系列教材编委会
2019 年 6 月

面向新工科专业建设计算机系列教材编委会

主　任：

　　张尧学　清华大学计算机科学与技术系教授　中国工程院院士/教育部高等学校
　　　　　　软件工程专业教学指导委员会主任委员

副主任：

　　陈　刚　浙江大学计算机科学与技术学院　　　　　　　院长/教授
　　卢先和　清华大学出版社　　　　　　　　　　　　　　常务副总编辑、
　　　　　　　　　　　　　　　　　　　　　　　　　　　副社长/编审

委　员：

　　毕　胜　大连海事大学信息科学技术学院　　　　　　　院长/教授
　　蔡伯根　北京交通大学计算机与信息技术学院　　　　　院长/教授
　　陈　兵　南京航空航天大学计算机科学与技术学院　　　院长/教授
　　成秀珍　山东大学计算机科学与技术学院　　　　　　　院长/教授
　　丁志军　同济大学计算机科学与技术系　　　　　　　　系主任/教授
　　董军宇　中国海洋大学信息科学与工程学部　　　　　　部长/教授
　　冯　丹　华中科技大学计算机学院　　　　　　　　　　院长/教授
　　冯立功　战略支援部队信息工程大学网络空间安全学院　院长/教授
　　高　英　华南理工大学计算机科学与工程学院　　　　　副院长/教授
　　桂小林　西安交通大学计算机科学与技术学院　　　　　教授
　　郭卫斌　华东理工大学信息科学与工程学院　　　　　　副院长/教授
　　郭文忠　福州大学数学与计算机科学学院　　　　　　　院长/教授
　　郭毅可　香港科技大学　　　　　　　　　　　　　　　副校长/教授
　　过敏意　上海交通大学计算机科学与工程系　　　　　　教授
　　胡瑞敏　西安电子科技大学网络与信息安全学院　　　　院长/教授
　　黄河燕　北京理工大学计算机学院　　　　　　　　　　院长/教授
　　雷蕴奇　厦门大学计算机科学系　　　　　　　　　　　教授
　　李凡长　苏州大学计算机科学与技术学院　　　　　　　院长/教授
　　李克秋　天津大学计算机科学与技术学院　　　　　　　院长/教授
　　李肯立　湖南大学　　　　　　　　　　　　　　　　　副校长/教授
　　李向阳　中国科学技术大学计算机科学与技术学院　　　执行院长/教授
　　梁荣华　浙江工业大学计算机科学与技术学院　　　　　执行院长/教授
　　刘延飞　火箭军工程大学基础部　　　　　　　　　　　副主任/教授
　　陆建峰　南京理工大学计算机科学与工程学院　　　　　副院长/教授
　　罗军舟　东南大学计算机科学与工程学院　　　　　　　教授
　　吕建成　四川大学计算机学院(软件学院)　　　　　　　院长/教授
　　吕卫锋　北京航空航天大学　　　　　　　　　　　　　副校长/教授

马志新	兰州大学信息科学与工程学院	副院长/教授
毛晓光	国防科技大学计算机学院	副院长/教授
明　仲	深圳大学计算机与软件学院	院长/教授
彭进业	西北大学信息科学与技术学院	院长/教授
钱德沛	北京航空航天大学计算机学院	中国科学院院士/教授
申恒涛	电子科技大学计算机科学与工程学院	院长/教授
苏　森	北京邮电大学	副校长/教授
汪　萌	合肥工业大学	副校长/教授
王长波	华东师范大学计算机科学与软件工程学院	常务副院长/教授
王劲松	天津理工大学计算机科学与工程学院	院长/教授
王良民	江苏大学计算机科学与通信工程学院	院长/教授
王　泉	西安电子科技大学	副校长/教授
王晓阳	复旦大学计算机科学技术学院	教授
王　义	东北大学计算机科学与工程学院	院长/教授
魏晓辉	吉林大学计算机科学与技术学院	教授
文继荣	中国人民大学信息学院	院长/教授
翁　健	暨南大学	副校长/教授
吴　迪	中山大学计算机学院	副院长/教授
吴　卿	杭州电子科技大学	教授
武永卫	清华大学计算机科学与技术系	副主任/教授
肖国强	西南大学计算机与信息科学学院	院长/教授
熊盛武	武汉理工大学计算机科学与技术学院	院长/教授
徐　伟	陆军工程大学指挥控制工程学院	院长/副教授
杨　鉴	云南大学信息学院	教授
杨　燕	西南交通大学信息科学与技术学院	副院长/教授
杨　震	北京工业大学信息学部	副主任/教授
姚　力	北京师范大学人工智能学院	执行院长/教授
叶保留	河海大学计算机与信息学院	院长/教授
印桂生	哈尔滨工程大学计算机科学与技术学院	院长/教授
袁晓洁	南开大学计算机学院	院长/教授
张春元	国防科技大学计算机学院	教授
张　强	大连理工大学计算机科学与技术学院	院长/教授
张清华	重庆邮电大学计算机科学与技术学院	执行院长/教授
张艳宁	西北工业大学	副校长/教授
赵建平	长春理工大学计算机科学技术学院	院长/教授
郑新奇	中国地质大学(北京)信息工程学院	院长/教授
仲　红	安徽大学计算机科学与技术学院	院长/教授
周　勇	中国矿业大学计算机科学与技术学院	院长/教授
周志华	南京大学计算机科学与技术系	系主任/教授
邹北骥	中南大学计算机学院	教授

秘书长：

白立军	清华大学出版社	副编审

前言

"网络工程"专业以"网络工程建设"为核心,培养学生掌握计算机网络工程技术的基本理论、方法与应用,运用所学知识与技能去分析和解决相关的实际问题,成为能够在计算机网络工程及相关领域中从事各类网络系统及计算机通信系统的研究、教学、设计、开发等工作的高级科技人才。

网络的应用几乎覆盖了人们生活、工作、娱乐等各个区域,培养学生认识网络,提高学生的网络应用能力具有重要的意义。目前,全国绝大多数高校都开设有"网络工程"专业,而计算机网络与通信网络(包括有线网络和无线网络)的结合是本校区别于其他高校"网络工程"专业的显著特色。

本书以弘扬伟大建党精神,传承红色基因,落实"立德树人"根本任务,推进党的二十大精神和习近平新时代中国特色社会主义思想进教材为目标。明确为谁培养人、培养什么样的人、怎样培养人的问题。围绕网络工程建设的核心,依托《网络工程》(第3版)一书,结合计算机网络的实际应用,培养学生网络项目建设以及解决常见网络故障问题的能力。引导青年学子努力成为堪当民族复兴重任的时代新人。

本书主要有以下特色:①实验目的明确,针对性强,不仅能很好地验证理论,还有助于提高学生对网络拓扑的设计和知识的应用能力;②实验内容与实际应用紧密结合,有助于解决实际问题;③实验内容设计合理,由浅入深,对于学习和巩固知识有很大作用;④实用性强,使用范围广,既可作为网络工程专业的教材,也可作为通信专业以及网络爱好者学习的指导教程;⑤为了更好地实践学习,对于典型的实验或重要的配置内容,有视频参考,帮助读者理解;⑥书中有完整的实验拓扑图,并且在基础部分详细介绍了实验环境的搭建方法,能够有效指导学生实践。

本书由西安交通大学城市学院的杨恩宁、李联宁、胡朋和许大炜编写及统稿;网络工程专业的学生周芊芊、王丹丹、贺文婷、宋文迪、寇晨旭参与实验的验证;网络与新媒体专业的学生张亮和张雅楠负责视频文件的处理。感谢西安交通大学陆丽娜、缪相林教授为本书提出的很多很好的建议。

限于编者的水平,书中难免有不妥之处,恳请使用本书的教师和学生批评指正,提出宝贵意见和建议。

编 者

2023年5月

目录

基 础 篇

实验一　认识网络 ··· 3
    一、实验目的 ·· 3
    二、实验环境 ·· 3
    三、实验内容及步骤 ·· 3
    四、实验过程 ·· 3
    五、实验总结 ·· 6
    六、实验思考题 ··· 6

实验二　网线的制作 ··· 7
    一、实验目的 ·· 7
    二、实验环境 ·· 7
    三、实验内容及步骤 ·· 7
    四、实验过程 ·· 7
    五、实验总结 ·· 9
    六、实验思考题 ··· 9

实验三　常用网络命令的使用 ··· 10
    一、实验目的 ·· 10
    二、实验环境 ·· 10
    三、实验内容及步骤 ·· 10
    四、实验过程 ·· 10
    五、实验总结 ·· 17
    六、实验思考题 ··· 17

实验四　快速以太网的构建 ··· 18
    一、实验目的 ·· 18
    二、实验环境 ·· 18

三、实验内容及步骤 …… 18
　　四、实验过程 …… 18
　　五、实验总结 …… 30
　　六、实验思考题 …… 30

实验五　网络设备 IOS 在 CLI 模式下的基本使用 …… 31
　　一、实验目的 …… 31
　　二、实验环境 …… 31
　　三、实验内容及步骤 …… 31
　　四、实验过程 …… 31
　　五、实验总结 …… 39
　　六、实验思考题 …… 39

实验六　交换机的常用配置与管理 …… 40
　　一、实验目的 …… 40
　　二、实验环境 …… 40
　　三、实验内容及步骤 …… 40
　　四、实验过程 …… 40
　　五、实验总结 …… 45
　　六、实验思考题 …… 45

实验七　基于交换机端口的 VLAN 划分 …… 46
　　一、实验目的 …… 46
　　二、实验环境 …… 46
　　三、实验内容及步骤 …… 46
　　四、实验过程 …… 46
　　五、实验总结 …… 52
　　六、实验思考题 …… 52

实验八　VLAN 间路由设计 …… 53
　　一、实验目的 …… 53
　　二、实验环境 …… 53
　　三、实验内容及步骤 …… 53
　　四、实验过程 …… 53
　　五、实验总结 …… 58
　　六、实验思考题 …… 58

实验九　CDP 的配置与验证 …… 59
　　一、实验目的 …… 59

二、实验环境 ……………………………………………………………………… 59

　　三、实验内容及步骤 ……………………………………………………………… 59

　　四、实验过程 ……………………………………………………………………… 59

　　五、实验总结 ……………………………………………………………………… 63

　　六、实验思考题 …………………………………………………………………… 63

实验十　静态路由的配置与验证 …………………………………………………… 64

　　一、实验目的 ……………………………………………………………………… 64

　　二、实验环境 ……………………………………………………………………… 64

　　三、实验内容及步骤 ……………………………………………………………… 64

　　四、实验过程 ……………………………………………………………………… 64

　　五、实验总结 ……………………………………………………………………… 69

　　六、实验思考题 …………………………………………………………………… 69

实验十一　默认路由的配置与验证 ………………………………………………… 70

　　一、实验目的 ……………………………………………………………………… 70

　　二、实验环境 ……………………………………………………………………… 70

　　三、实验内容及步骤 ……………………………………………………………… 70

　　四、实验过程 ……………………………………………………………………… 70

　　五、实验总结 ……………………………………………………………………… 75

　　六、实验思考题 …………………………………………………………………… 75

实验十二　RIP 的配置与验证 ……………………………………………………… 76

　　一、实验目的 ……………………………………………………………………… 76

　　二、实验环境 ……………………………………………………………………… 76

　　三、实验内容及步骤 ……………………………………………………………… 76

　　四、实验过程 ……………………………………………………………………… 77

　　五、实验总结 ……………………………………………………………………… 81

　　六、实验思考题 …………………………………………………………………… 82

实验十三　网络应用服务平台的构建与测试 ……………………………………… 83

　　一、实验目的 ……………………………………………………………………… 83

　　二、实验环境 ……………………………………………………………………… 83

　　三、实验内容及步骤 ……………………………………………………………… 83

　　四、实验过程 ……………………………………………………………………… 83

　　五、实验总结 ……………………………………………………………………… 90

　　六、实验思考题 …………………………………………………………………… 90

实验十四 小型无线网络的设计与配置 …… 91

一、实验目的 …… 91

二、实验环境 …… 91

三、实验内容及步骤 …… 91

四、实验过程 …… 92

五、实验总结 …… 105

六、实验思考题 …… 105

实验十五 网络打印服务的配置 …… 106

一、实验目的 …… 106

二、实验环境 …… 106

三、实验内容及步骤 …… 106

四、实验过程 …… 106

五、实验总结 …… 116

六、实验思考题 …… 116

技能提升篇

实验十六 生成树协议的原理与验证 …… 119

一、实验目的 …… 119

二、实验环境 …… 119

三、实验内容及步骤 …… 119

四、实验过程 …… 120

五、实验总结 …… 124

六、实验思考题 …… 124

实验十七 OSPF 协议的配置与验证 …… 125

一、实验目的 …… 125

二、实验环境 …… 125

三、实验内容及步骤 …… 125

四、实验过程 …… 126

五、实验总结 …… 131

六、实验思考题 …… 131

实验十八 BGP 的配置与验证 …… 132

一、实验目的 …… 132

二、实验环境 …… 132

三、实验内容及步骤 ·· 132
　　四、实验过程 ··· 133
　　五、实验总结 ··· 137
　　六、实验思考题 ··· 137

实验十九　VRRP 的配置与验证 ··· 138
　　一、实验目的 ··· 138
　　二、实验环境 ··· 139
　　三、实验内容及步骤 ·· 139
　　四、实验过程 ··· 139
　　五、实验总结 ··· 142
　　六、实验思考题 ··· 142

实验二十　OSPF 协议的高级配置与验证 ·· 143
　　一、实验目的 ··· 143
　　二、实验环境 ··· 143
　　三、实验内容及步骤 ·· 143
　　四、实验过程 ··· 144
　　五、实验总结 ··· 153
　　六、实验思考题 ··· 153

实验二十一　IS-IS 协议的配置与验证 ·· 154
　　一、实验目的 ··· 154
　　二、实验环境 ··· 154
　　三、实验内容及步骤 ·· 155
　　四、实验过程 ··· 155
　　五、实验总结 ··· 160
　　六、实验思考题 ··· 160

实验二十二　交换机端口安全配置与验证 ·· 161
　　一、实验目的 ··· 161
　　二、实验环境 ··· 161
　　三、实验内容及步骤 ·· 161
　　四、实验过程 ··· 161
　　五、实验总结 ··· 164
　　六、实验思考题 ··· 164

实验二十三　访问控制列表的配置与验证 ·· 165
　　一、实验目的 ··· 165

二、实验环境 ··· 165
　　三、实验内容及步骤 ··· 165
　　四、实验过程 ··· 166
　　五、实验总结 ··· 170
　　六、实验思考题 ·· 170

实验二十四　防火墙安全策略的设计与配置 ······························ 171
　　一、实验目的 ··· 171
　　二、实验环境 ··· 171
　　三、实验内容及步骤 ··· 171
　　四、实验过程 ··· 171
　　五、实验总结 ··· 179
　　六、实验思考题 ·· 180

实验二十五　典型企业网络的设计与实现 ································· 182
　　一、实验目的 ··· 182
　　二、实验环境 ··· 183
　　三、实验内容及步骤 ··· 183
　　四、实验过程 ··· 183
　　五、实验总结 ··· 189
　　六、实验思考题 ·· 189

参考文献 ·· 190

基础篇

实验一 认识网络

一、实验目的

（1）了解常用网络设备的名称、规格、性能等参数。
（2）了解常用网络结构及连接方法。
（3）参观校园网络中心，了解校园网的架构。

二、实验环境

网卡、交换机、路由器、防火墙、网络实验室、校园网络中心。

三、实验内容及步骤

1. 网络适配器

通过教师对网络适配器的讲解，掌握网络适配器的用途，熟悉网络适配器的参数及性能。

2. 交换机

熟悉交换机在网络中的作用，了解交换机的工作原理，熟悉交换机的性能和参数信息。

3. 路由器

认识路由器，熟悉路由器在网络中的作用，了解路由器的工作原理，熟悉路由器的分类、参数等信息。

4. 防火墙

通过教师对防火墙的讲解，掌握防火墙的概念和作用，了解网络安全相关知识。

5. 网络实验室

参观网络实验室，了解网络实验室的结构，熟悉网络实验体系，了解网络工程专业方向的培养计划。

6. 校园网

通过对本校校园网络中心的参观学习，掌握本校校园网的结构、网络三层架构等情况，进一步理解网络拓扑结构、网络中的设备应用以及网络的管理和运行等知识。

四、实验过程

1. 网络适配器

网络适配器就是网卡，它可以使计算机接入网络，一般固化在主板上，负责把用户要传递的数据转换为网络上其他设备能够识别的格式，再通过网络传输介质传输。按照使用方向可以分为普通网卡、服务器网卡和无线网卡，它的主要技术参数为带宽、总线方式、电气接口方式等，网卡要正常使用还需安装相应的驱动程序。

2．交换机

交换机(switch)是一种能在通信系统中完成信息交换功能的设备，主要功能包括物理编址、错误校验、网络拓扑结构实现、帧序列校验以及流控，其主要工作原理有以下几点。

（1）交换机根据收到数据帧中的源 MAC 地址建立该地址同交换机端口的映射，并将其写入 MAC 地址表中。

（2）交换机将数据帧中的目的 MAC 地址同已建立的 MAC 地址表进行比较，以决定由哪个端口进行转发。

（3）如数据帧中的目的 MAC 地址不在 MAC 地址表中，则向所有端口转发。这一过程称为泛洪。

（4）广播帧和组播帧向所有的端口转发。

交换机是数据链路层的设备，只关注 MAC 地址，不涉及 IP 地址。

3．路由器

路由器(router)是一种计算机网络互联设备，它能将数据打包，根据信道的情况自动选择和设定路由，以最佳路径将数据传送至目的地。它是互联网络的枢纽，工作在 OSI 模型的第三层——网络层。

4．防火墙

防火墙(firewall)是一种由软件和硬件设备组合而成、在内部网和外部网之间、专用网与公共网之间构造的保护屏障，使 Internet 与 Intranet 之间建立起一个安全网关，从而保护内部网免受非法用户的侵入。防火墙主要由服务访问规则、验证工具、包过滤和应用网关 4 部分组成。

5．参观网络实验室

由实验教师带领学生参观网络实验室，并讲解网络实验室的构成，所拥有的网络设备硬件环境，以及网络实验体系。图 1-1 所示的实验平台专门针对网络实验定制，可以满足 CCNA 级别的各类实验。

图 1-1 实验环境示例

6．参观校园网络中心

由教师带队，前往校园网络中心进行参观学习，由校园网络中心的教师讲解校园网的架构，深入学习校园网的拓扑结构，使用的网络设备，软件环境，校园网的运行和管理等知识，每一位学生都要做好详细的记录。随后再到各楼，参观校园网的汇聚间以及网络的终端，掌握网络的实际分布和应用。校园网络拓扑如图 1-2 所示。

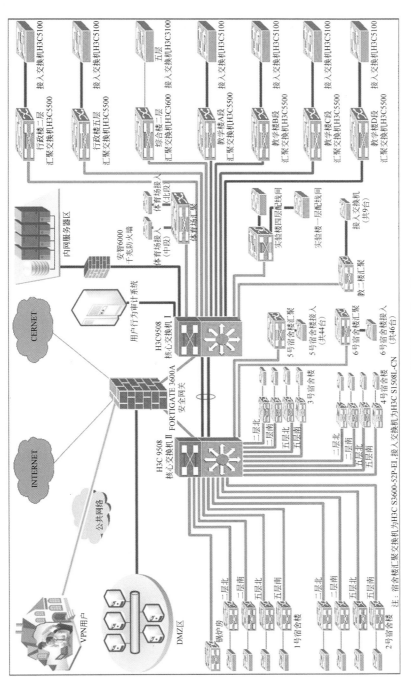

图 1-2 校园网络拓扑示例

五、实验总结

　　本实验让学生熟悉网卡、交换机和路由器,掌握网络中常用设备的原理和应用,参观网络实验室和校园网,掌握实验环境、专业培养体系和校园网络结构。在参观过程中,要求学生要遵守纪律和各种规章制度,认真听讲,并做好详细的记录,特别是一些设备的规格、参数等更要仔细记录,并写在实验报告中。

六、实验思考题

　　(1) 写出在参观过程中看到的网络设备的名称及相关参数。
　　(2) 绘制本校校园网的网络拓扑图。

实验二 网线的制作

目前网络主流的传输介质是双绞线和光纤,而在以双绞线为传输介质的网络中,跳线的制作与测试非常重要。对于小型网络而言,网线连接着集线设备与计算机;对于大中型网络而言,网线既连接着信息插座与计算机,也连接着集线设备与跳线设备及跳线板。总之,无论如何,网线的制作与测试是网络管理员必须掌握的入门级技能。

一、实验目的

(1) 了解网络中常用的传输介质。
(2) 掌握双绞线的定义标准及制作规范。
(3) 掌握 RJ-45 头的制作以及网线连通性的测试方法。
(4) 熟悉网络的互联方式。

二、实验环境

RJ-45 头若干、双绞线 1~2m、RJ-45 压线钳一把、测试仪一套。

三、实验内容及步骤

(1) 熟悉双绞线的分类和结构。
(2) 直通 UTP 电缆的制作。
(3) 交叉 UTP 电缆的制作。
(4) 网线连通性的测试。

四、实验过程

1. 双绞线

双绞线是局域网中最基本的传输介质,由不同颜色的 4 对 8 芯线组成,每两根线按照一定的标准缠绕在一起,一般缠绕越密其抗干扰能力就越强。与其他传输介质相比,双绞线在传输距离、信道宽度和数据传输速度等方面均受到一定的限制,但其价格较为低廉。双绞线分为非屏蔽双绞线和屏蔽双绞线,最大传输距离 100m,如果要加大网络范围,可安装中继器,进行信号的再生放大,但一般最多安装 4 个,使网络最大范围达到 500m。

2. 双绞线的线序标准

1) EIA/TIA-568-A 标准

EIA/TIA-568-A 简称 T568A,其双绞线的排列顺序为:绿白、绿、橙白、蓝、蓝白、橙、棕白、棕。依次插入 RJ-45 头的 1~8 号线槽中,如表 2-1 所示。

表 2-1 T568A 标准

线槽	1	2	3	4	5	6	7	8
颜色	绿白	绿	橙白	蓝	蓝白	橙	棕白	棕

2) EIA/TIA-568-B 标准

EIA/TIA-568-B 简称 T568B,其双绞线的排列顺序为:橙白、橙、绿白、蓝、蓝白、绿、棕白、棕。依次插入 RJ-45 头的 1~8 号线槽中,如表 2-2 所示。

表 2-2 T568B 标准

线槽	1	2	3	4	5	6	7	8
颜色	橙白	橙	绿白	蓝	蓝白	绿	棕白	棕

如果双绞线的两端均采用同一标准(如 T568B),则称这根双绞线为直通线。直通线是一种用得最多的连接方式,能用于异种网络设备间的连接,如计算机与集线器的连接、集线器与路由器的连接。

如果双绞线的两端采用不同的连接标准(如一端用 T568A,另一端用 T568B),则称这根双绞线为跳接(交叉)线。跳接线能用于同种类型设备的连接,如计算机与计算机的直联、集线器与集线器的级联。需要注意的是,有些集线器(或交换机)本身带有"级联端口",当用某一集线器的"普通端口"与另一集线器的"级联端口"相连时,因"级联端口"内部已经做了"跳接"处理,所以这时只能用"直通"双绞线来完成其连接,如图 2-1 和图 2-2 所示。

图 2-1 直通 UTP 线缆

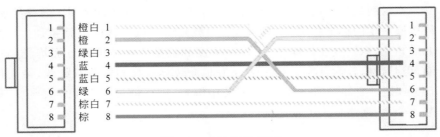

图 2-2 交叉 UTP 线缆

3. 双绞线 RJ-45 头的制作

(1) 从头部开始去掉双绞线 20mm 左右的外部套层,并将 8 根导线理直。

(2) 确定是直通线还是交叉线式,然后按照对应关系将双绞线中的线色按顺序排列,不要有差错。

(3) 将非屏蔽 5 类双绞线的 RJ-45 接头点处切齐,并且使裸露部分保持在 12mm 左右。

(4) 将双绞线整齐地插入 RJ-45 接头中(塑料扣的一面朝下,开口朝右)。

(5) 用 RJ-45 压线钳将接头压实即可。

(6) 注意在双绞线压接处不能拧、撕,防止有断线的伤痕;使用 RJ-45 压线钳时,要压实,不能有松动。

4. 网线的测试

将做好的双绞线两端的 RJ-45 头分别插入测试仪两端,打开测试仪电源开关检测制作是否正确。如果测试仪的 8 个指示灯按从上到下的顺序循环呈现绿灯,则说明双绞线制作正确;如果 8 个指示灯中有的呈现绿灯,有的呈现红灯,则说明双绞线线序出现问题;如果 4 个指示灯中有的呈现绿灯,有的不亮,则说明双绞线存在接触不良的问题。

五、实验总结

本实验让学生掌握制作 RJ-45 头的方法,若学生制作并不熟练,成功率不高,则还需要加强练习。

1. 易犯的错误

(1) 剥线时将铜线剪断。

(2) 线缆没有整理整齐就插入接头,可能使某些铜线并未插入正确的插槽。

(3) 线缆插入过短,导致铜线并未与铜片紧密接触。

2. 故障排除

测试仪的指示灯不亮:测试仪使用的电池没电,线缆断裂或 RJ-45 头制作不当。

插头接触不良:网卡、集线器、测线器的 RJ-45 连接接口的 8 个接点对应,有 8 条铜线,插入时铜线内缩,插入次数多了以后,铜线的弹性降低。

六、实验思考题

(1) 双绞线中的线缆为何要成对地绞在一起,其作用是什么?

(2) 网线测试仪除了测试线缆的连通性外,还能进行其他有关线缆性能的测试吗?

实验三 常用网络命令的使用

常用网络命令

网络命令是基于计算机网络的运行、管理和维护，在 DOS 环境下运行，是 DOS 操作系统的命令。现在的计算机系统虽然都是可视化、图形化的界面，但是为了更好地管理计算机，在 Windows 的操作系统下都集成了 MS-DOS 环境，用户可以通过系统中的命令提示符窗口打开 DOS 环境，通过网络命令，对计算机网络进行管理和维护。

一、实验目的

（1）了解 DOS 系统。
（2）掌握 Windows 系统下 DOS 命令的使用方法。
（3）掌握 ipconfig、arp、ping 等命令的操作使用。

二、实验环境

计算机（Windows 7 操作系统）局域网。

三、实验内容及步骤

（1）DOS 系统和 DOS 命令的认识。
（2）ipconfig 命令的使用。
（3）arp 命令的使用。
（4）ping 命令的操作使用。
（5）tracert 命令的使用。

四、实验过程

1. DOS 系统和 DOS 命令的认识

DOS 是磁盘操作系统的缩写，是个人计算机上的一类操作系统，且在操作系统中占有举足轻重的地位。DOS 家族包括 MS-DOS、PC-DOS、DR-DOS、Free-DOS、PTS-DOS、ROM-DOS、JM-OS 等，其中以 MS-DOS 最为著名。自微软图形界面操作系统 Windows NT 问世以来，DOS 就是以后台程序的形式出现的，可以通过在命令提示符窗口输入 CMD 进入运行，界面如图 3-1 所示。

2. ipconfig 命令

（1）功能：获得主机配置信息，包括 IP 地址、子网掩码和默认网关。

图 3-1 DOS 命令窗口

（2）如果不带任何参数选项，那么，它已经配置了接口 IP 地址、子网掩码和默认网关值，如图 3-2 所示。

图 3-2 ipconfig 命令使用

（3）ipconfig/all：当使用参数 all 时，能显示 DNS 和 WINS 服务器的配置信息，以及内置于本地网卡中的物理地址（MAC），如图 3-3 所示。

（4）ipconfig/release 和 ipconfig/renew。这两项属于附加选项，只能在向 DHCP 服务器租用其 IP 地址的计算机上起作用。如果输入 ipconfig/release，那么所有接口的租用 IP 地址便重新交付给 DHCP 服务器（归还 IP 地址）；如果输入 ipconfig/renew，那么本地计算机便设法与 DHCP 服务器取得联系，并租用一个 IP 地址。请注意，大多数情况下网卡将被重新赋予与以前所赋予的相同的 IP 地址。

3. arp 命令

1）功能

ARP（地址解析协议）是根据 IP 地址获取物理地址的一个 TCP/IP，可以保证通信的顺利进行。可在高速缓冲区建立 ARP 表，并通过 arp 命令进行查看、添加和删除，实现 ARP 表的管理、维护。

2）ARP 表分类

（1）动态表项（dynamic）：随时间推移自动添加和删除。

（2）静态表项（static）：一直存在，直到人为删除或重新启动。

3）arp 命令操作

（1）arp -a：显示高速 cache 中的 ARP 表，如图 3-4 所示。

图 3-3 ipconfig/all 命令使用

图 3-4 查看 ARP 表

（2）arp -s：添加 ARP 静态表项。格式为 arp -s inet_addr ether_addr，即在 ARP 缓存中添加项，将 IP 地址 inet_addr 和物理地址 ether_addr 关联。手动添加 IP 地址为 192.168.32.60，MAC 地址为 00-d0-09-f0-33-71 等 ARP 静态表项信息，如图 3-5 所示。

（3）arp -d：删除 ARP 表项。格式为 arp -d inet_addr，即删除由 inet_addr 指定的项，如图 3-6 所示。

如果要删除所有 ARP 表信息，可使用 arp -d * 或 arp -d，如图 3-7 所示。

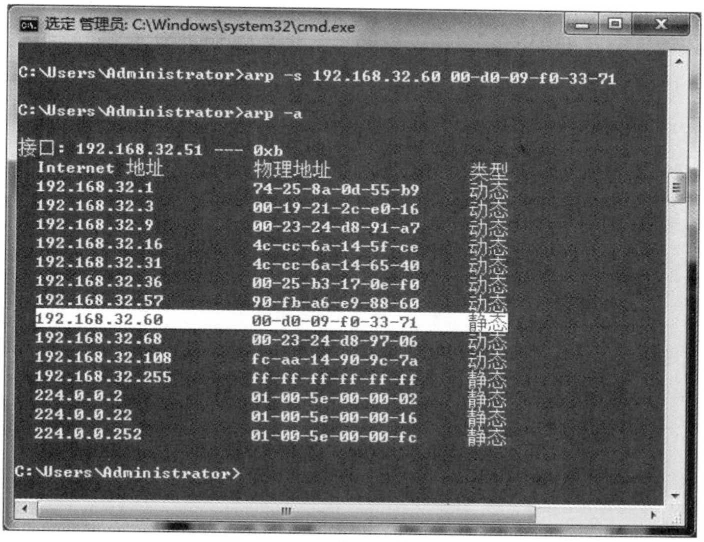

图 3-5 手动添加 ARP 表

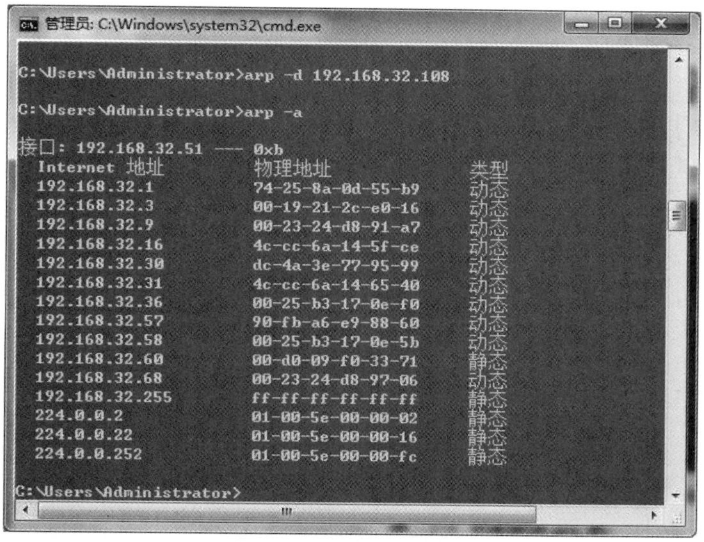

图 3-6 删除 ARP 表项

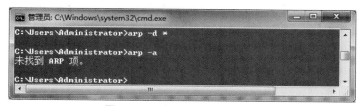

图 3-7 删除所有 ARP 表信息

4. ping 命令

ping 命令的全称叫作 ping.exe,是微软公司提供的用来进行网络连接测试的工具,能够帮助网络工程师准确、快速地判断出网络故障,是网络运行维护中最常用的命令之一。

1) ping 命令形式

具体形式为 ping [-t] [-a] [-n count] [-l size] [-f] [-i TTL] [-v TOS] [-r count] [-s count] [[-j host-list] | [-k host-list]] [-w timeout] 目的主机/IP 地址,如图 3-8 所示。

图 3-8 ping 地址 192.168.32.1

2) 连续发送 ping 探测报文

具体形式为 ping -t 192.168.32.1,如图 3-9 所示。

图 3-9 连续发送 ping 包

此时,可按组合键 Ctrl+Break 查看统计信息,按快捷键 Ctrl+C 结束命令。

3) 自选数据长度的 ping 探测报文

具体形式为 ping 目的主机/IP 地址 -l size,如图 3-10 所示。

4) 修改 ping 命令的请求超时时间

具体形式为 ping -w 目的主机/IP 地址,表示指定等待每个回送应答的超时时间,单位

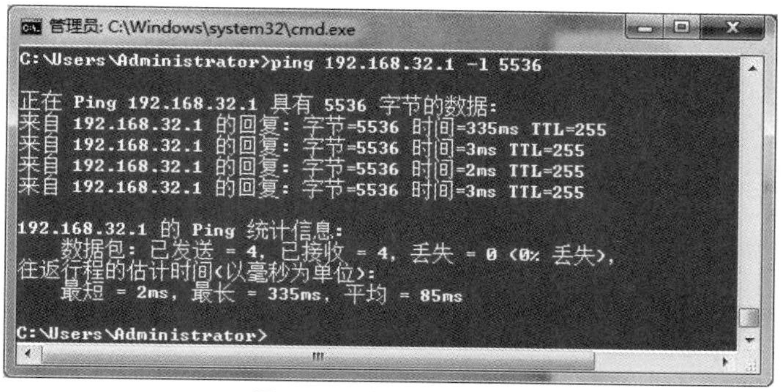

图 3-10　自选数据长度

为毫秒(ms)，默认值为 1000ms，如图 3-11 所示。

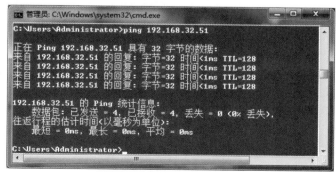

图 3-11　修改请求超时时间

5）ping 命令应用

假设目前实验用机器通信协议配置如下。

本机 IP：192.168.32.51。

子网掩码：255.255.255.0。

默认网关：192.168.32.1。

DNS 服务器的 IP 地址：192.168.11.16。

（1）验证网卡工作是否正常。打开 MS-DOS 环境，在提示符后输入"ping 192.168.32.51"（本机 IP 地址），按 Enter 键运行，若出现图 3-12 所示的信息，则说明网卡工作正常；若出现图 3-13 所示的信息，则说明网卡工作不正常。

图 3-12　网卡工作正常

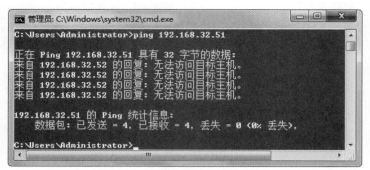

图 3-13　网卡工作不正常

（2）验证局域网是否畅通。打开 MS-DOS 环境，在提示符后输入"ping 192.168.32.101"（局域网中的一台计算机），按 Enter 键运行，若出现图 3-14 所示的信息，则说明网卡工作正常；否则，说明网卡工作不正常。

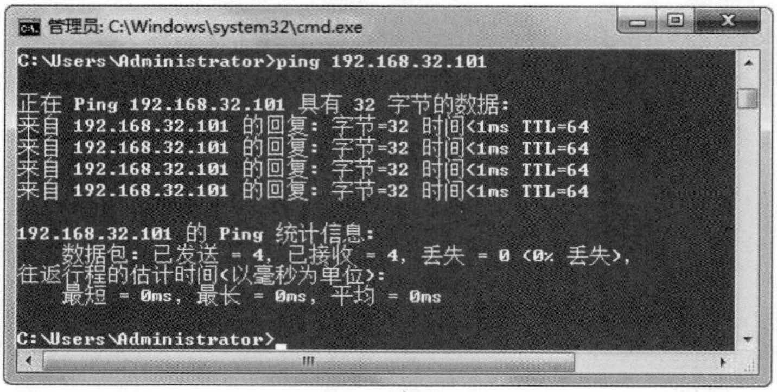

图 3-14　ping 局域网中的计算机

（3）验证 DNS 配置正确与否。打开 MS-DOS 环境，在提示符后输入任一域名（如 www.hao123.com），看其是否能被解析成一个 IP 地址。如输入"ping www.hao123.com"，若出现图 3-15 所示的信息，说明 DNS 服务器配置正确；若出现图 3-16 所示的信息，说明 DNS 配置错误。

图 3-15　得到 DNS 解析地址

图 3-16　DNS 配置错误

5. tracert 命令

（1）Tracert（跟踪路由）是路由跟踪实用程序，用于确定 IP 数据包访问目标所采取的路径。tracert 命令使用 IP 生存时间（TTL）字段和 ICMP 错误消息来确定从一个主机到网络上其他主机的路由。

（2）tracert 命令的功能：可以探测源结点到目的结点之间数据报文经过的路径。

（3）tracert 命令的常用格式如下：

```
tracert ip_adress
tracert host_name
```

接下来对某网站进行路由跟踪，如图 3-17 所示。

图 3-17　路由跟踪

五、实验总结

本实验让学生掌握 ipconfig、arp、ping 等常用网络命令的操作使用方法，方便学生对网络配置信息进行查看，以及对网络故障进行检测判断和处理。

六、实验思考题

（1）一般情况下如何利用 ping 命令检查网络故障？

（2）tracert 命令后面能否添加 IP 地址，这种形式和加域名有什么区别？

实验四 快速以太网的构建

快速以太网(100Mbps)采用的是 CSMA/CD 访问控制方法,符合 IEEE 802.3 标准。快速以太网与原来在 100Mbps 带宽下工作的 FDDI 相比具有许多的优点,最主要体现在快速以太网技术可以有效地保障用户在布线等基础设施上的投资,它支持 3、4、5 类双绞线以及光纤的连接。快速以太网在功能上也有不足之处,这也是以太网技术的不足,快速以太网仍是基于 CSMA/CD 技术的,当网络负载较重时,会造成效率的降低,当然,这可以使用交换技术来弥补。100Mbps 快速以太网标准又分为 100BASE-TX、100BASE-FX、100BASE-T4 这 3 个子类。

一、实验目的

(1) 掌握网络适配器的作用,安装配置、连通性测试等的方法。
(2) 掌握组建快速以太网的技术与方法。
(3) 认识网络拓扑,搭建简单的网络拓扑结构。
(4) 掌握模拟器中计算机的配置和网络连通性的测试方法。

二、实验环境

网络适配器、交换机、网线、计算机、服务器、测试仪。

三、实验内容及步骤

(1) 熟悉快速以太网组建的步骤。
(2) 安装网络适配器、驱动程序以及配置 TCP/IP。
(3) 选择通信线缆,掌握直通线和交叉线的使用规则。
(4) 熟悉模拟器 Cisco Packet Tracer 的使用。
(5) 快速以太网的构建。
(6) 网络的测试。

四、实验过程

1. 快速以太网组建的步骤

(1) 安装以太网适配器。
(2) 将计算机接入网络。
(3) 安装配置网络适配器驱动程序。

(4) 安装和配置 TCP/IP。

(5) 多交换机互连：使用直通 UTP 线缆，如图 4-1 所示。

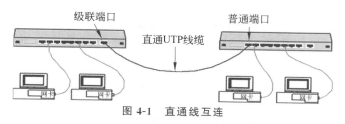

图 4-1　直通线互连

(6) 多交换机互连：使用交叉 UTP 线缆，如图 4-2 所示。

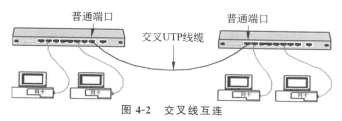

图 4-2　交叉线互连

在网络中采用交叉线和直通线连接，是因为以前的设备端口收发线序是固定的。例如，两台交换机相连，发送端口使用的都是 1、2 芯，接收端口使用的都是 3、6 芯。这样，交换机的连接就需用交叉线，以保证两台交换机之间发送和接收数据对应。如果交换机和计算机连接，计算机发送端口是 3、6 芯，接收端口是 1、2 芯，就需要用到直通线连接。

注意：现在，通信设备的 RJ-45 接口基本都能自适应，遇到网线不匹配的情况，可以自动翻转端口的接收和发送，因此现在基本都用直通线。

2. 安装网络适配器、驱动程序以及配置 TCP/IP

(1) 驱动程序的主要功能：实现网络操作系统上层程序与网络适配器的接口。

(2) 网络适配器驱动程序因网络适配器和操作系统的不同而不同。

(3) 驱动程序一般随同网络适配器一起发售，但有些常用的驱动程序也可以在操作系统安装盘中找到；操作系统会自动安装好驱动程序。

(4) 在支持"即插即用"的操作系统中使用"即插即用"型网络适配器，不需要手工安装和配置（中断号的分配等），只要安装好驱动程序即可。

(5) 手工安装：可根据系统界面提示进行操作。

3. TCP/IP 的安装和配置

说明：本安装过程是在 Windows Server 系统下进行的，仅供参考。现实操作时以真实环境为准，细节上虽有不同，但整体思路不变。

1) 通过控制面板中的网络选项添加 TCP/IP

在控制面板界面双击"网络"选项，如果在网络组件类型对话框中没有发现 TCP/IP，则需要单击"添加"按钮，然后双击对话框中的协议选项，如图 4-3 所示。

在左边厂商中选择 Microsoft 选项，右边网络协议里选择 TCP/IP，然后单击"确定"按钮，如图 4-4 所示。

图 4-3 选择需要安装的网络组件

图 4-4 添加 TCP/IP

2）TCP/IP 的配置

单击"本地连接"→"属性"按钮，单击"Internet 协议（TCP/IP）"选项，如图 4-5 所示，再单击"属性"按钮，就会出现"Internet 协议（TCP/IP）属性"对话框，选择"使用下面的 IP 地址(S)："单选按钮，在 IP 地址框中输入相应的 IP 地址和子网掩码，然后单击"确定"按钮返回，如图 4-6 所示。

图 4-5 "本地连接 属性"对话框

4. 通信线缆的使用规则

双绞线一共八根线，八根线的布线规则是 1236 线有用，4578 线闲置。其中，1、2 为传输

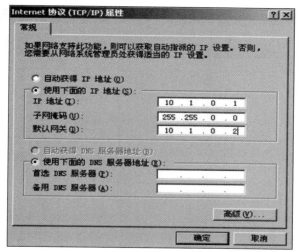

图 4-6　TCP/IP 配置

线、3、6 为接收线。网线有两种做法,一种是直通线,另一种是交叉线。

关于网线的制作,"实验二"中已有详细介绍,此处不再赘述。

需要注意的是,计算机与路由器相连用交叉线。

5. 模拟器 Cisco Packet Tracer 的使用

Cisco Packet Tracer(以下简称 Cisco PT)是由 Cisco 公司发布的一个辅助学习工具,为 Cisco 网络课程的初学者设计、配置、排除网络故障提供了网络模拟环境。用户可以在软件的图形用户界面上建立网络拓扑,并且软件可提供数据包在网络中行进的详细处理过程,用户利用软件提供的记录可以观察网络的实时运行情况,还可以学习网络拓扑的构建,常见路由交换设备的典型配置、调试,从而锻炼故障排查能力。

Cisco PT 6.0 基本操作

1) Cisco PT 的获取和安装

(1) Cisco PT 模拟器的获取。可通过 Cisco 官方网站或第三方软件下载。

(2) Cisco PT 模拟器的安装。该模拟软件的功能强大,安装简单,只需要双击可执行文件 Cisco Packet Tracer.exe,根据安装向导的提示安装即可。

2) Cisco PT 使用界面功能介绍

模拟器安装成功后,会在桌面建立相应的快捷方式,双击该快捷方式即可启动该软件,打开 Cisco PT 的主界面,如图 4-7 所示。

如果用户英语能力不是很强,也可以使用该模拟软件的汉化版,其在功能上与英文版软件没有区别,同样可以完成相应的实验,达到学习的效果,成为网络高手。

(1) Cisco PT 的主要功能特点如下。

① 模拟硬件设备,与真实设备差不多,面板、模块和功能一样,操作简单,容易上手,很适合初学者学习和使用。

② 可模拟设备种类多,如 PC,集线器,二层、三层交换机,路由器等,可实现的实验范围广。

③ 模拟器包含的工具、功能丰富,可直接完成网络拓扑的设计和实现。

④ 支持数据报文分析,可以更好地掌握网络通信原理知识以及进行网络故障的判断。

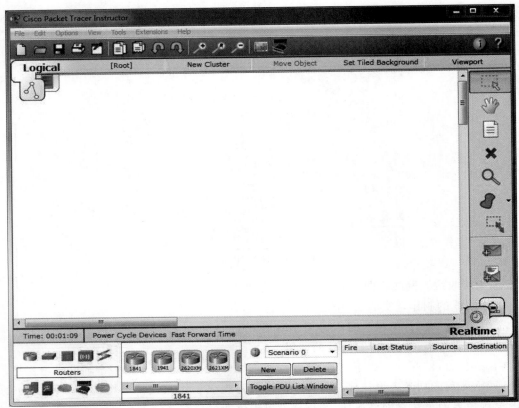

图 4-7　Cisco PT 主界面

⑤ 模拟器有无线系统和广域网模块，可以进行相关实验操作，提升学生对无线知识和广域网知识的理解和应用能力。

(2) Cisco PT 的界面构成介绍如下。

Cisco PT 的界面简单，最上面是标题栏，下面一栏是主菜单，再往下就是常用工具栏，中间空白区域是操作窗口（搭建实验环境的主阵地），最下面主要是网络设备选择区域和传输介质选择区域，界面右侧是常用的编辑区域，如图 4-8 所示。

Cisco PT 模拟器功能强大，界面内容丰富，操作简单，还包含 Realtime Mode（实时模式）和 Simulation Mode（模拟模式）。

下面对 Cisco PT 界面中的部分功能进行简单介绍。

① 网络设备。

Routers（路由器，快捷键为 Ctrl+Alt+R）。Cisco PT 提供了 1841、2811、2911 等多种路由器设备，路由器配置界面有 Physical、Config 和 CLI 这 3 个标签，可以通过 Physical 标签进行模块的装载。路由器上有自带的接口，也预留着一些空槽，用户可以在空槽上安装相应模块，方法很简单。只需要用鼠标左键选中模块，然后拖曳至空槽位置即可。但是切记一定要关闭电源，否则无法完成模块的装载，一般默认设备的电源是开启的，显示为绿色，如图 4-9 所示。

在 Config 选项卡中可以直观地看到路由的名称、接口配置、路由协议信息等，当然也可

实验四　快速以太网的构建

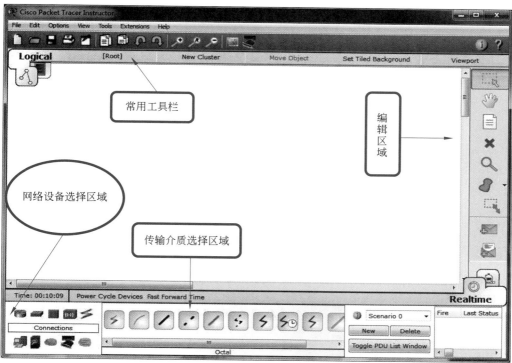

图 4-8　Cisco PT 界面介绍

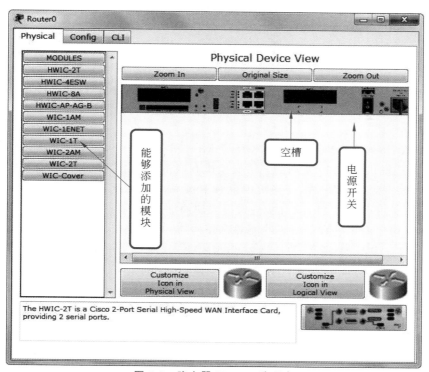

图 4-9　路由器 Physical 选项卡

以通过该选项卡进行上述信息的配置。配置相应信息的同时，可以发现，系统会自动在 Config 选项卡的下面生成相应的命令，但是这种配置模式一般不推荐，因为在真实的设备中不存在该项，除非用户对命令行模式比较熟悉，而这项功能只是为了节省配置的时间。Config 选项卡的具体信息如图 4-10 所示。

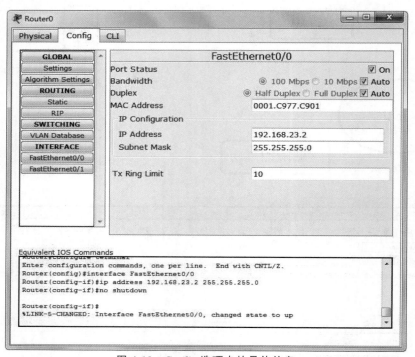

图 4-10 Config 选项卡的具体信息

CLI 是路由器设备的命令行模式，通过该模式可以对路由器进行相关配置，与真实设备没有区别，关于 CLI 命令行的使用会在实验六做详细的讲解和应用。CLI 命令行模式如图 4-11 所示。

Switches（交换机，快捷键为 Ctrl＋Alt＋S）。除了 2950、2960 等系列的二层交换机，Cisco PT 还提供了 3560 系列的三层交换机和网桥，大大扩展了实验的范围并提高了实验的可操作性。和路由器一样，交换机中也有 Physical、Config 和 CLI 这 3 个标签，只不过只有极少数设备可以通过 Physical 标签进行模块添加，其他设备的物理接口都是固化的，只可进行查看操作。Config 和 CLI 两个标签和路由器的作用差不多，可以对比学习。

Hubs（分线器、集线器，快捷键为 Ctrl＋Alt＋U）。Cisco PT 提供了中继器、集线器和分离器，目前已经很少使用，只是用作部分原理知识的验证。

Wireless Devices（无线设备，快捷键为 Ctrl＋Alt＋W）。Cisco PT 6.0 模拟器还提供了部分无线设备，满足用户进行无线网络学习的需求，方便无线网络的搭建和测试，有利于网络专业知识的拓展。

Connections（通信链路，快捷键为 Ctrl＋Alt＋O）。Cisco PT 6.0 模拟器提供了多种通信线缆，用于网络中设备间的互连。

End Devices（终端设备，快捷键为 Ctrl＋Alt＋V）。Cisco PT 6.0 模拟器提供了电

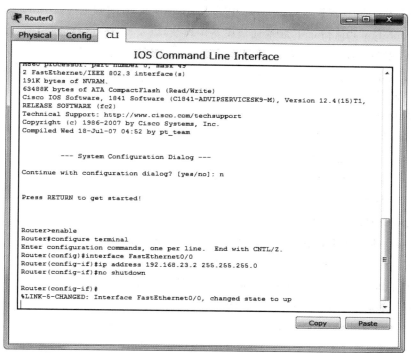

图 4-11 CLI 命令行模式

话、计算机、服务器等多种终端设备，方便用户进行网络的构建和测试。

Security(安全设备，快捷键为 Ctrl+Alt+C)。该版本的模拟器中提供的是思科 ASA5505 防火墙。

WAN Emulation(WAN 仿真，快捷键为 Ctrl+Alt+N)。Cisco PT 6.0 模拟器对广域网进行仿真，方便用户搭建网络模型，进行广域网相关的实验，以开拓知识结构。

Cisco PT 6.0 模拟器还提供了 Custom Made Devices(定制设备，快捷键为 Ctrl+Alt+T)和 Multiuser Connection(多用户连接器，快捷键为 Ctrl+Alt+N)。

② 传输介质。

Automatically Choose Connection Type(自动选择链路)，它可根据用户选择的网络设备自动匹配传输介质。这项功能是万能的，但是不建议使用，除非真的不知道设备间应该用什么线缆。当然它也受制于实验环境，因为真实的环境中设备的互连就必须选择正确的线缆。

Console 线，也叫配置线，是专门用于网络设备配置的，将计算机的串口和设备的 Console 端口相连接，通过计算机的超级终端对网络设备进行配置。

Copper Straight-Through(直通线)，该线缆的两端采用相同的线序标准，主要用于网络中不同设备的互连，例如交换机与计算机的互连、交换机与路由器的互连等。

Copper Cross-Over(交叉线)，两端采用不同的线序标准，若一端用 568A，则另外一端用 568B，该线缆主要用于相同设备的互连，例如交换机与交换机、路由器与计算机互连等。

Fiber(光纤)，用于设备的千兆接口连接，一般用于骨干网络，传输距离远、速率高。

Cisco PT 模拟器还提供了 Phone(电话线)、Coaxial(同轴电缆)、Serial DCE(Serial Data Communication Equipment,数据通信设备串口线)以及 Serial DTE(Serial Data Terminal Equipment,数据终端设备串口线)。后两种线缆主要用于路由器并通过 Serial 端口进行连接,如果选择前者,则首先连接路由器的接口就是 DCE 端,必须配置时钟,后选中的路由器的接口就默认为 DTE 端;如果选中的是后者,则需要人为定义 DCE 端和 DTE 端。

③ 设备编辑功能。

　　Select(快捷键为 Esc):选定/取消,拖曳选定的设备。

　　Move layout(快捷键为 M):移动画布。

　　Place note(快捷键为 N):放置书签/贴便条。

　　Delete(快捷键为 Delete):删除。

　　Inspect(快捷键为 I):插入。

　　Resize Shape(快捷键为 Alt+R):调整形状大小。

　　Add Simple PDU(快捷键为 P):添加简单协议数据单元。

　　Add Complex PDU(快捷键为 C):添加复杂协议数据单元。

3) 使用 Cisco PT 6.0 构建一个简单的网络拓扑

通过对 Cisco PT 6.0 软件的学习,完成如图 4-12 所示的网络拓扑。

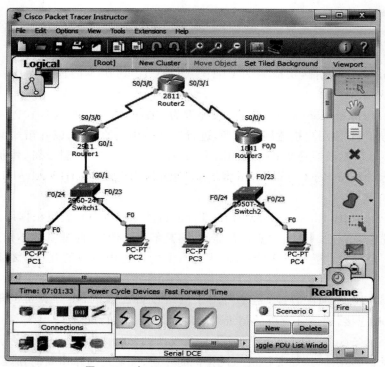

图 4-12　在 Cisco PT 6.0 中构建网络拓扑

具体操作方法如下。

(1) 如图 4-12 所示选择相应设备,在网络设备选择区域找到所需设备,然后按住鼠标左键将其拖曳到编辑区域。

(2)双击路由器,打开 Physical 选项卡,关闭设备电源开关,装载串口模块 WIC-2T。
(3)选择 Serial DCE 线缆,按图 4-12 连接路由器。
(4)注意交换机与计算机以及路由器之间用直通线连接。
(5)IP 地址的配置信息如表 4-1 所示。

表 4-1 IP 地址配置信息表

序 号	设 备	IP 地 址	子 网 掩 码
1	PC1	192.168.32.50	255.255.255.0
2	PC2	192.168.32.51	255.255.255.0
3	PC3	192.168.32.52	255.255.255.0
4	PC4	192.168.32.53	255.255.255.0

6. 快速以太网的组建

1)计算机的直连

在现实的学习工作中,有时需要将两台计算机互连,实现资源共享,这就是最简单的快速以太网,其具体组建过程如下。

快速以太网的组建

(1)设备选择:计算机 2 台。
(2)端口连接与配置:选择交叉线,将两台计算机的网卡连接起来,如图 4-13 所示。

图 4-13 计算机的直连

配置计算机 PC0 的 IP 地址,单击计算机 PC0 图标,选择 Desktop→IP Configuration 选项,打开 IP 地址配置对话框,如图 4-14 所示进行配置。

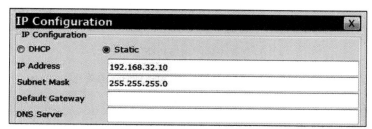

图 4-14 计算机 PC0 的 IP 地址配置

按照同样的方式将计算机 PC1 的 IP 地址设置为 192.168.32.20,子网掩码设置为 255.255.255.0。

(3)网络测试:测试 PC1 和 PC0 的连通性,打开 PC1 的 Command Prompt 命令窗口,输入 ping 192.168.32.10,按 Enter 键后,若得到图 4-15 所示的信息,就说明 PC1 和 PC0 的网络连通正常。

2)多计算机网络的构建

在实际的应用中,还需要将多台计算机互连,形成较大规模的局域网络。此时,就需要

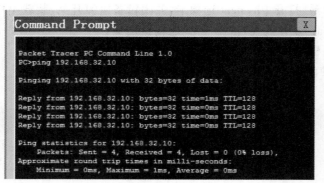

图 4-15 连通性测试

借助网络设备——交换机,让所有的计算机都和交换机互连,进而达到所有设备互连的目的,具体组建过程如下。

(1) 设备选择:2950-24 型号的交换机 1 台,计算机 4 台。

(2) 端口连接与配置:搭建如图 4-16 所示的网络拓扑,实现设备互连,由于计算机和交换机属于不同设备,因此需选用直通线,具体端口连接及 IP 地址的配置信息如表 4-2 所示。

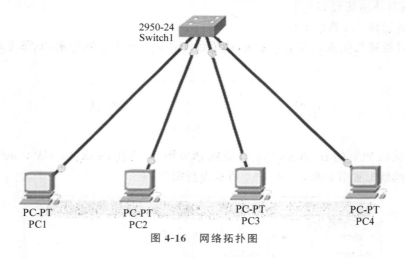

图 4-16 网络拓扑图

表 4-2 端口连接及 IP 地址配置信息表

序号	设备名称	本机端口	交换机端口	IP 地址	子网掩码
1	PC1	FastEthernet0	F0/1	192.168.32.10	255.255.255.0
2	PC2	FastEthernet0	F0/2	192.168.32.20	255.255.255.0
3	PC3	FastEthernet0	F0/3	192.168.32.30	255.255.255.0
4	PC4	FastEthernet0	F0/4	192.168.32.40	255.255.255.0

需要注意的是,FastEthernet 为快速以太网口类型,其缩写方式比较灵活,可以表示为 f、fa 等,且对于英文字母的大小写不做严格要求,本书中的 GigabitEthernet Serial、Ethernet 等接口类型同理。

（3）网络测试：选择两台计算机，利用 ping 命令进行连通性测试，如果能得到图 4-17 所示的信息，说明网络连接正常，可实现数据通信，否则网络故障。

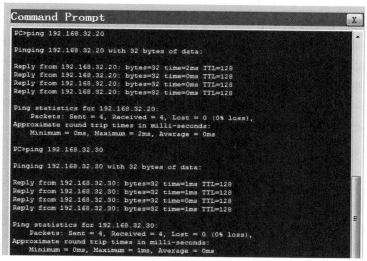

图 4-17　网络测试

3）跨交换机网络的构建

（1）设备选择：选择两台 2950T-24 型号交换机 SW1 和 SW2 以及 4 台计算机。

（2）设备连接与配置：SW1 与计算机 PC1 和 PC2 相连，SW2 与 PC3 和 PC4 相连，计算机和交换机都使用直通线连接；SW1 和 SW2 通过交叉线将千兆以太口 GigabitEthernet0/1 相连，拓扑如图 4-18 所示，端口和 IP 地址的设置仍采用表 4-2 中的信息。

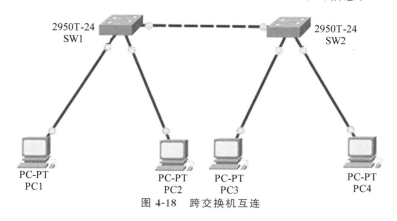

图 4-18　跨交换机互连

（3）网络测试：首先用 PC1 ping PC2，PC3 ping PC4，如果得到图 4-15 所示的信息，说明 SW1 和 SW2 内部的网络畅通；再用 PC1 ping PC3，如果得到图 4-17 所示的信息，说明跨交换机的网络连通正常。

一般网络连通性的测试也可以通过以下两种方式来实现。

（1）观察交换机和网络适配器状态指示灯的变化。将网线的一端插入交换机的 RJ-45 端口，另一端插入计算机的网络适配器的 RJ-45 端口，这时观察端口边的指示灯变化情况。如果指示灯亮，表明网络连接正常；如果指示灯亮，并且开始闪烁，表明这时有数据传输。

(2) 利用高层命令和软件(如 ping 命令等)。

五、实验总结

本次实践学习,让学生熟悉 Cisco Packet Tracer 工具软件的界面、功能和应用,掌握网络拓扑的构建、设备的选型、设备模块的装载、传输介质的选择、网络的互联等知识。结合网络理论知识,能够利用该模拟器搭建实验环境,验证网络原理知识,构建网络拓扑,配置网络设备,实现网络通信,采集网络数据,分析数据包信息,排查网络故障以及掌握快速以太网组建的要领,掌握网络适配器的作用及应用,配置 TCP/IP,掌握如何选择网络设备和通信线缆,熟悉网络连通的各种测试方法。

六、实验思考题

(1) 连接在同一网段上的计算机,如果有两台或两台以上的计算机使用相同的 IP 地址,会出现什么情况?

(2) 如果在一个网络中,某台计算机 ping 另外一台主机不通,而 ping 其他 IP 地址的主机均能通,则故障的可能原因有哪些?

实验五 网络设备 IOS 在 CLI 模式下的基本使用

一、实验目的

(1) 熟悉路由器的基本原理和结构。
(2) 熟悉路由器 CLI 的各种管理模式。
(3) 熟悉路由器 CLI 的各种编辑命令。
(4) 掌握路由器的 IOS 基本命令。
(5) 学会查看路由器的有关信息。

路由器 CLI 的使用与 IOS 基本命令

二、实验环境

Cisco Packet Tracer 6.0 模拟器(计算机、配置线缆、服务器)。

三、实验内容及步骤

(1) 学习路由器的内部结构和工作原理。
(2) 了解路由器的管理模式。
(3) 在模拟器中搭建网络拓扑。
(4) 学习路由器的 CLI 命令行模式和 IOS 基本命令。
(5) 对网络设备进行配置,实现网络连通。
(6) 通过 show 命令查看路由器的配置信息。

四、实验过程

1. 路由器的内部结构和工作原理

把数据从一个地方传送到另一个地方的过程叫作路由,而路由就是由路由器来完成的。路由器是一种连接多个网络或网段的网络设备,它是网络层的设备,具有判断网络地址和进行路径选择的功能。路由选择是靠路由表来实现的,路由表中有各种传输路径的数据信息,路由表信息可以由管理员手动设定,也可以动态获取。由系统管理员事先设置好固定的路由表称为静态(static)路由表,它不会随网络结构的改变而改变;路由器根据网络系统的运行情况而自动调整的路由表称为动态路由表。

1) 路由器的主要功能

(1) 网络互联:路由器支持各种局域网和广域网接口,主要用于互联局域网和广域网,实现不同网络间的互相通信。

(2) 数据处理：路由器提供包括分组过滤、分组转发、优先级、复用、加密、压缩和防火墙等功能。

(3) 网络管理：路由器提供包括配置管理、性能管理、容错管理和流量控制等功能。

2) 路由器的内部结构

路由器就是一台有特殊用途的计算机，但和计算机相比没有键盘、鼠标、硬盘、显示器等，除了有 CPU、内存、BOOT ROM 外还多了 NVRAM、FLASH 以及各种各样的接口。

(1) CPU：中央处理单元，和计算机一样，它是路由器的控制和运算部件。

(2) RAM/DRAM：内存，用于存储临时的运算数据，如路由表、ARP 表、快速交换缓存、缓冲数据包、数据队列、当前配置等，众所周知，RAM 中的数据在路由器断电后是会丢失的。

(3) FLASH：可擦除、可编程的 ROM，用于存放路由器的 IOS，FLASH 的可擦除特性允许用户更新、升级 IOS 而不用更换路由器内部的芯片，路由器断电后，FLASH 的内容不会丢失，FLASH 容量较大时，就可以存放多个 IOS 版本。

(4) NVRAM：非易失性 RAM，用于存放路由器的配置文件，路由器断电后，NVRAM 中的内容仍然可以保存。

(5) ROM：只读存储器，存储了路由器的开机诊断程序、引导程序和特殊版本的 IOS 软件，ROM 中软件升级时需要更换芯片。

(6) 接口（Interface）：用于网络连接，路由器就是通过这些接口与不同的网络进行连接的。

2. 路由器常用的管理模式

(1) 通过超级终端管理。通过计算机上的超级终端访问路由器的命令行界面 CLI，进而对路由器进行配置和管理。

(2) 使用远程 telnet 命令管理。用户可以借助网络通过 Telnet 客户端程序登录路由器，对路由器进行配置和管理。

(3) 使用支持 SNMP 的网络管理软件管理。路由器配置管理地址，运行网络管理软件，对设备进行配置管理。

3. 搭建实验拓扑

(1) 实验设备选择：1841 路由器 1 台、2950 交换机 1 台、计算机 3 台、服务器 1 台。

(2) 线缆选择：交叉线、直通线以及配置线。

(3) 构建网络拓扑，如图 5-1 所示。

4. 路由器的基本配置

1) 搭建配置环境

(1) 用配置线缆将计算机的 RS-232 接口与路由器的 Console 端口连接，搭建配置环境。

(2) 设置通信参数，如图 5-2 所示。

(3) 单击图 5-2 对话框中的 OK 按钮，打开终端窗口进行配置，如图 5-3 所示。

2) 路由器的 4 种模式

(1) 用户模式，该模式是路由器的第一个模式，在用户模式下可以查看路由器的版本信息，进行简单的测试，可用命令相当有限，提示符为 Router＞。

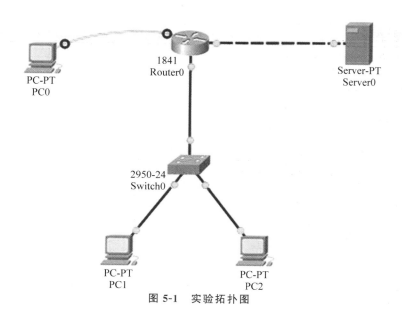

图 5-1　实验拓扑图

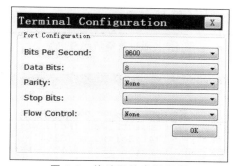

图 5-2　终端通信参数设置

（2）特权模式，这是用户模式的下一级配置模式，在该模式下可以查看路由器的配置信息，测试网络的连通性，管理配置文件等，提示符为 Router#。

（3）全局配置模式，属于特权模式的下一级配置模式，通过全局配置模式可以设置路由器的用户名、密码以及路由器的具体功能，提示符为 Router(config)#。

（4）端口配置模式，也叫接口模式，是全局配置模式的下一级配置模式，可以对路由器的接口进行具体的配置，提示符为 Router(config-if)#。

在路由器的配置过程中，可以使用 exit 命令返回上一级模式，也可以使用 end 命令直接退回到特权模式，CLI 配置界面支持命令简写、自动补齐以及获取帮助等功能。

3）路由器 CLI 配置模式以及 IOS 基本命令的使用

建立链接，打开超级终端，观察路由器的启动过程，在路由器的默认配置提示窗口输入命令 no，连按两次 Enter 键，进入路由器的用户模式，如图 5-4 所示。

（1）命令窗口中?的使用。在当前用户模式下输入?，就会显示在该模式下能执行的所有命令，可以使用户了解在该模式下能进行的操作，更好地帮助用户进行路由器的配置，如图 5-5 所示。

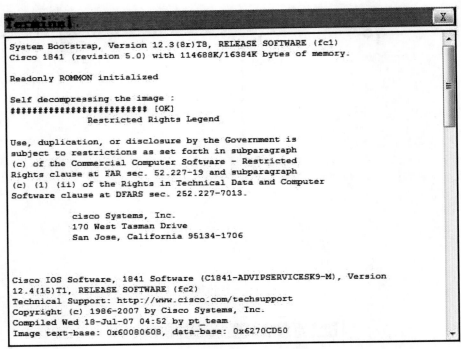

图 5-3　通过终端访问路由器

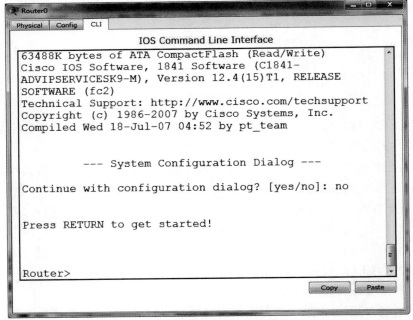

图 5-4　进入路由器的用户模式

（2）输入命令 enable，按 Enter 键进入路由器的特权模式，再输入?，按 Enter 键查看该模式下的可用命令，了解在该模式下能够对路由器进行的配置，如图 5-6 所示。

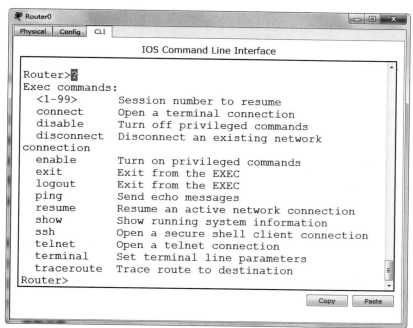

图 5-5　用户模式下的所有命令

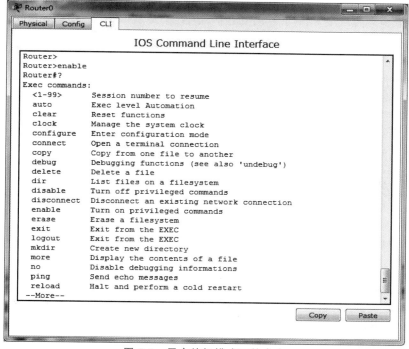

图 5-6　用户特权模式下的命令

若出现--More--则说明信息没有显示完全，按 Enter 键显示下一行，按 Space 键显示下一页，按其他键则退出。

（3）利用?帮助命令,实现对路由器时钟的配置,命令内容如下:

```
Router#cl?
clear clock
//路由器列出了当前模式下可以使用的以cl开头的所有命令
Router#clock
% Incomplete command.
//路由器提示命令输入不完整
Router#clock ?
set Set the time and date
//需要注意的是clock和?之间要有空格,否则将得到不同的结果。如果不加空格,路由器以为用
//户想列出以clock字母开头的命令,而不是想列出clock命令的子命令或参数
Router#clock set ?
hh:mm:ss Current Time
Router#clock set 11:36:00
% Incomplete command.
Router#clock set 11:36:00 ?
<1-31> Day of the month
MONTH Month of the year
Router#clock set 11:36:00 12 ?
MONTH Month of the year
//以上多次使用?帮助命令,获得了clock命令的格式
Router#clock set 11:36:00 12 08
% Invalid input detected at '^' marker.
//路由器提示输入了无效的参数,并用^指示错误所在
Router#clock set 11:36:00 12 august
% Incomplete command.
Router#clock set 11:36:00 12 august 2018
Router#show clock
11:36:03.149 UTC Tue Aug 12 2018
//到此成功配置了路由器的时钟,通常如果命令成功,路由器不会有任何提示
```

（4）在特权模式下输入conf,按Tab键,系统会自动将命令configure补齐,再输入t,按Tab键,将自动补齐命令terminal,按Enter键后系统将进入全局配置模式,示例如下:

```
Router#conf+【tab】
Router#configure t+【tab】
Router#configure terminal
Enter configuration commands, one per line. End with CNTL/Z.
Router(config)#
```

注意补齐的前提是,在该模式下,输入的部分能够唯一标识命令。例如,通过命令重新启动路由器,但是在该模式下有命令reload和resume,输入re时系统是无法自动补齐的,必须输入rel,再按Tab键才可补齐。

(5) 对路由器进行名称和密码的设置,命令内容如下:

```
Router(config)#hostname jsj001
jsj001(config)#enable password cisco
jsj001(config)#
```

(6) 接口设置,打开端口,配置端口 IP 地址,命令内容如下:

```
jsj001(config)#interface fastEthernet 0/0
jsj001(config-if)#no shutdown
%LINK-5-CHANGED: Interface FastEthernet0/0, changed state to up
%LINEPROTO-5-UPDOWN: Line protocol on Interface FastEthernet0/0, changed state to up
jsj001(config-if)#ip address 192.168.32.1 255.255.255.0
jsj001(config-if)#exit
jsj001(config)#interface fastEthernet 0/1
jsj001(config-if)#no shutdown
%LINK-5-CHANGED: Interface FastEthernet0/1, changed state to up
%LINEPROTO-5-UPDOWN: Line protocol on Interface FastEthernet0/1, changed state to up
jsj001(config-if)#ip address 192.168.30.1 255.255.255.0
jsj001 (config-if)#exit
```

经过以上配置,物理端口被打开,可以直观地看到线路上的灯变亮。

(7) 保存配置信息,返回至特权模式,使用 copy 命令,命令内容如下:

```
jsj001 (config-if)#end
jsj001#copy running-config startup-config
Destination filename [startup-config]?
Building configuration...
[OK]
jsj001#
```

返回命令有 exit 和 end 两种,前者是返回上一级模式,后者是直接返回到特权模式。保存是将配置文件信息保存到启动文件中,需要二次确认,没有执行保存,路由器一旦重启,所有配置信息将丢失。

(8) 查看配置信息,在特权模式下,通过 show 命令查看路由器的配置信息,命令内容如下:

```
jsj001#show running-config
Building configuration...
Current configuration : 612 bytes
version 12.4
no service timestamps log datetime msec
no service timestamps debug datetime msec
```

```
no service password-encryption
hostname jsj001
enable password cisco
ip cef
no ipv6 cef
spanning-tree mode pvst
interface FastEthernet0/0
ip address 192.168.32.1 255.255.255.0
duplex auto
speed auto
interface FastEthernet0/1
ip address 192.168.30.1 255.255.255.0
duplex auto
speed auto
interface Vlan1
no ip address
shutdown
ip classless
ip flow-export version 9
no cdp run
line con 0
line aux 0
line vty 0 4
login
end
```

（9）网络测试。配置计算机和服务器的 IP 地址信息，如表 5-1 所示。

表 5-1 IP 地址配置信息表

序 号	设 备	IP 地 址	子网掩码	网 关
1	PC1	192.168.30.50	255.255.255.0	192.168.30.1
2	PC2	192.168.30.51	255.255.255.0	192.168.30.1
3	Server	192.168.32.52	255.255.255.0	192.168.32.1

打开 PC1 命令提示符窗口，分别 ping PC2 和 Server，得到如下信息：

```
PC>ping 192.168.30.51
Pinging 192.168.30.51 with 32 bytes of data:
Reply from 192.168.30.51: bytes=32 time=0ms TTL=128
Reply from 192.168.30.51: bytes=32 time=0ms TTL=128
Reply from 192.168.30.51: bytes=32 time=0ms TTL=128
Reply from 192.168.30.51: bytes=32 time=1ms TTL=128
Ping statistics for 192.168.30.51:
Packets: Sent = 4, Received = 4, Lost = 0 (0% loss),
```

```
Approximate round trip times in milli-seconds:
Minimum = 0ms, Maximum = 1ms, Average = 0ms
PC>ping 192.168.32.52
Pinging 192.168.32.52 with 32 bytes of data:
Request timed out.
Reply from 192.168.32.52: bytes=32 time=0ms TTL=127
Reply from 192.168.32.52: bytes=32 time=0ms TTL=127
Reply from 192.168.32.52: bytes=32 time=0ms TTL=127
Ping statistics for 192.168.32.52:
Packets: Sent = 4, Received = 3, Lost = 1 (25% loss),
Approximate round trip times in milli-seconds:
Minimum = 0ms, Maximum = 0ms, Average = 0ms
```

以上信息说明通过对路由器的配置,网络已经连通,可以在计算机与服务器之间的网络进行通信。

五、实验总结

本实验可以让学生更清楚地认识路由器,了解路由器的基本结构以及工作原理,掌握路由器的配置管理方法,学会利用计算机通过超级终端对路由器进行基本的配置,熟悉路由器的用户模式、特权模式、全局配置模式以及端口配置模式,掌握 enable、exit、end、copy unning-config starting-config、password、hostname、show 等命令的使用,实现网络拓扑,使得网络中的结点能够互相访问。

六、实验思考题

(1) 路由器的两个接口能不能配置相同网段的地址?
(2) 网关的作用是什么,计算机不配网关能不能访问服务器?

实验六 交换机的常用配置与管理

交换机的基本配置

一、实验目的

（1）熟悉交换机的基本原理和结构。
（2）掌握交换机的基本配置方法。
（3）掌握远程 telnet 命令管理交换机的方法。
（4）掌握交换机配置信息的备份方法。

二、实验环境

Cisco Packet Tracer 模拟器（计算机、交换机、服务器）。

三、实验内容及步骤

（1）了解交换机的工作原理。
（2）了解交换机的管理模式。
（3）在模拟器中搭建网络拓扑。
（4）对交换机进行基本配置。
（5）实现交换机的远程管理。
（6）通过 show 命令查看交换机的配置信息。
（7）对配置信息进行备份。

四、实验过程

1. 交换机的工作原理

交换机工作于数据链路层，位于 OSI 参考模型的第二层。它拥有一条很高带宽的背部总线和内部交换矩阵，所有的端口都挂接在这条背部总线上。交换机的 CPU 会在每个端口成功连接时，通过将 MAC 地址和端口对应，形成一张 MAC 表。在后续的通信中，发往该 MAC 地址的数据包将仅送往其对应的端口，若目的 MAC 地址不存在，就会广播到所有的端口，接收端口回应后，交换机会"学习"新的 MAC 地址，并加入 MAC 地址表中。交换机可用于划分数据链路层广播，即冲突域；但它不能划分网络层广播，即广播域。

2. 交换机常用的管理模式

（1）通过超级终端管理。通过计算机上的超级终端访问交换机的命令行界面 CLI，进而对交换机进行配置和管理。

（2）使用远程 telnet 命令管理。用户可以借助网络通过 Telnet 客户端程序登录交换机，对交换机进行配置和管理。

（3）使用支持 SNMP 的网络管理软件管理。

3. 搭建实验拓扑

（1）实验设备选择：2950-24 交换机 1 台、计算机 3 台、服务器 1 台。

（2）线缆选择：直通线和配置线。

（3）构建网络拓扑，如图 6-1 所示。

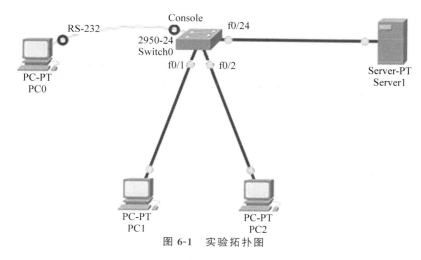

图 6-1　实验拓扑图

（4）IP 地址规划如表 6-1 所示。

表 6-1　IP 地址规划表

序号	设备	接口	IP 地址	子网掩码
1	PC1	FastEthernet0	172.16.0.10	255.255.0.0
2	PC2	FastEthernet0	172.16.0.20	255.255.0.0
3	Server1	FastEthernet 0/1	172.16.0.100	255.255.0.0
4	Switch	Vlan 1	172.16.0.1	255.255.0.0

4. 交换机的基本配置

1）搭建配置环境

（1）用配置线缆将计算机 PC0 的 RS-232 接口和交换机的 Console 端口连接，搭建配置环境。

（2）设置通信参数，如图 6-2 所示。

（3）单击 OK 按钮即可打开终端窗口，可以看到交换机的启动信息，如图 6-3 所示。

（4）如表 6-1 所示，配置 PC1、PC2 以及 Server1 的 IP 地址信息。

2）交换机的 6 种模式

（1）Switch>：用户模式，它是交换机的第一个模式，在用户模式下可以查看交换机的版本信息，进行简单的测试，可用命令相当有限。

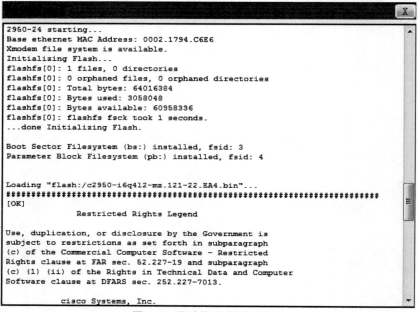

图 6-2　终端通信参数设置

图 6-3　通过终端访问交换机

（2）Switch♯：特权模式，它是用户模式的下一级配置模式，在该模式下可以查看交换机的配置信息，测试网络的连通性，管理配置文件等。

（3）Switch（config）♯：全局配置模式，属于特权模式的下一级配置模式，通过全局配置模式可以设置交换机的用户名、密码以及交换机的具体功能。

（4）Switch（config-if）♯：接口配置模式，是全局配置模式的下一级配置模式，可以对交换机的接口进行具体的配置。

（5）Switch（config-vlan）♯：VLAN 配置模式，也属于全局配置模式的下一级配置模式，主要进行交换机中各种 VLAN 的基本配置。

（6）Switch（config-line）♯：链路配置模式，对链路进行基本配置，主要应用于实现设备的远程管理。

在交换机的配置过程中，可以使用 exit 命令返回上一级模式，也可以使用 end 命令直接退回到特权模式，CLI 配置界面支持命令简写、自动补齐以及获取帮助等功能。

3）交换机的基本配置
（1）配置主机名，命令内容如下：

```
Switch>enable
Switch#configure terminal
Enter configuration commands, one per line. End with CNTL/Z.
Switch(config)#hostname SW1
```

（2）配置密码，命令内容如下：

```
SW1(config)#enable secret 123
```

（3）接口基本配置。交换机的以太网接口默认都是开启的，对于交换机的以太网口可以配置其双工模式、传输速率、接口模式等，命令内容如下：

```
SW1(config)#interface f0/1
SW1(config-if)#duplex ?
auto Enable AUTO duplex configuration
full Force full duplex operation
half Force half-duplex operation
//命令 duplex 用来配置接口的双工模式,full 表示全双工,half 表示半双工,auto 表示自动检
//测双工的模式
SW1(config-if)#speed ?
10   Force 10 Mbps operation
100  Force 100 Mbps operation
1000 Force 1000 Mbps operation
auto Enable AUTO speed configuration
//命令 speed 用来配置交换机的接口速率,10 表示 10Mbps,100 表示 100Mbps,1000 表示
//1000Mbps,auto 表示自动检测接口速率
```

5. 交换机远程管理
1）配置管理地址
交换机也允许被 telnet，这时需要在交换机上配置一个 IP 地址，这个地址是在 VLAN 接口上配置的，命令内容如下：

```
SW1(config)#int vlan 1
SW1(config-if)#ip address 172.16.0.1 255.255.0.0
SW1(config-if)#no shutdown
%LINK-5-CHANGED: Interface Vlan1, changed state to up
%LINEPROTO-5-UPDOWN: Line protocol on Interface Vlan1, changed state to up
SW1(config)#ip default-gateway 172.16.0.254
```

交换机默认所有端口都属于 VLAN 1，VLAN 1 默认为关闭状态。启动 VLAN 1，并配置相应的管理地址，就可以直接通过网络中的计算机进行 telnet 该地址来访问交换机。

2)配置远程 telnet

对交换机配置远程 telnet,命令内容如下:

```
SW1(config)#line vty 0 15
SW1(config-line)#password cisco123
SW1(config-line)#login
```

3)远程管理测试

打开 PC1 的 MS-DOS 环境,输入命令 telnet 172.16.0.1,输入链路密码,验证成功后进入 SW1 的用户模式,进入交换机的特权模式前也要进行密码验证,该密码为前面配置的设备密码(secret),如图 6-4 所示。

图 6-4 在 PC1 上远程登录 SW1

4)保存配置

保存上述配置,命令内容如下:

```
SW1#copy running-config startup-config
Destination filename [startup-config]?
Building configuration...
[OK]
```

6. 查看配置信息

在全局模式下,输入 show running-config 命令查看交换机的配置信息,命令内容如下:

```
SW1#show running-config
Building configuration...
Current configuration : 1117 bytes
version 12.1
no service timestamps log datetime msec
no service timestamps debug datetime msec
no service password-encryption
hostname SW1
enable secret 5 $1$mERr$3HhIgMGBA/9qNmgzccuxv0
spanning-tree mode pvst
interface FastEthernet0/1
duplexauto
speed 100
...
interface Vlan1
ip address 172.16.0.1 255.255.0.0
line vty 0 4
password cisco123
login
line vty 5 15
password cisco123
```

```
login
end
```

查看交换机的配置信息,会看到 secret 密码信息是一段乱码,这是因为采用了 MD5 算法加密,它的安全性更高。

7. 备份配置信息到 TFTP

首先配置 TFTP 的 IP 地址为 172.16.0.100,子网掩码为 255.255.0.0。然后对交换机 SW1 进行备份操作的配置,命令内容如下:

```
SW1#copy running-config tftp:
Address or name of remote host []? 172.16.0.100
Destination filename [SW1-confg]?peizhi
Writing running-config....!!
[OK - 1093 bytes]
1093 bytes copied in 3.012 secs (362 bytes/sec)
```

备份完成后,打开服务器 Server1 的配置界面,在 Services 选项卡的 SERVICES 选项中可以看到备份文件 peizhi,如图 6-5 所示。

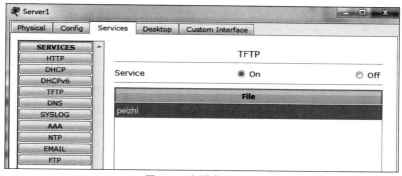

图 6-5　查看备份结果

五、实验总结

本实验可以让学生更清楚地认识交换机,了解交换机的基本结构以及工作原理,掌握交换机的配置、管理方法,学会利用计算机超级终端以及远程 telnet 命令对交换机进行配置和管理,熟悉交换机的用户模式、特权模式、全局配置模式以及各种端口配置模式,掌握基本命令的使用,实现了对交换机接口双工模式、速率的配置,掌握了对交换机进行管理地址、远程控制、数据备份等操作。

六、实验思考题

(1) 同时配置交换机的 password 密码和 secret 密码,哪一个会生效?

(2) 本实验中配置的 172.16.0.254 是什么地址,有什么作用?

实验七 基于交换机端口的 VLAN 划分

VLAN 划分

VLAN(virtual LAN,虚拟局域网)是交换机端口的逻辑组合,工作在 OSI 参考模型的第二层(数据链路层),VLAN 之间的通信是通过第三层的路由器来完成的。VLAN 的优点:①控制网络的广播,每一个 VLAN 是一个广播域,一个 VLAN 上的广播不会扩散到另一个 VLAN;②简化网络管理,当 VLAN 中的用户位置移动时,网络管理员只需设置几条命令,就可以完成网络的部署;③提高网络的安全性,VLAN 能控制广播,而 VLAN 之间不能直接通信。

一、实验目的

(1) 熟悉 VLAN 的概念和作用。
(2) 掌握 VLAN 的创建方法。
(3) 掌握基于交换机端口的 VLAN 划分方法。
(4) 掌握 trunk 链路的应用。

二、实验环境

Cisco Packet Tracer 6.0 模拟器(交换机、计算机、直通线、Console 配置线)。

三、实验内容及步骤

(1) 搭建实验拓扑。
(2) 进行 VLAN 的创建。
(3) 基于交换机端口的 VLAN 划分。
(4) 网络测试。
(5) trunk 链路应用。
(6) 查看端口配置信息。

四、实验过程

1. 实验拓扑

设备选择:2950 系列交换机 2 台、计算机 6 台。
通信线缆:交叉线 1 根,直通线 5 根。
端口连接:PC1、PC2 和 PC4 分别连接在 Switch1 的 f0/1、f0/2 和 f0/3 接口上,PC5、PC6、和 PC3 分别连接在 Switch2 的 f0/1、f0/2 和 f0/3 接口上,如图 7-1 所示。

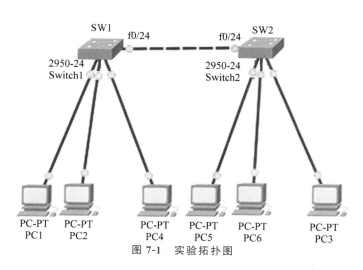

图 7-1 实验拓扑图

2. 地址规划

一般情况下,不同的 VLAN 表示不同的网络,所以要用不同网段的 IP 地址。为了体现 VLAN 划分前后的效果,本实验将 IP 地址设置为同一网段,如表 7-1 所示。

表 7-1 IP 地址规划表

序号	设备	接口	IP 地址	子网掩码
1	PC1	FastEthernet0	192.168.11.10	255.255.255.0
2	PC2	FastEthernet0	192.168.11.20	255.255.255.0
3	PC3	FastEthernet0	192.168.11.30	255.255.255.0
4	PC4	FastEthernet0	192.168.11.40	255.255.255.0
5	PC5	FastEthernet0	192.168.11.50	255.255.255.0
6	PC6	FastEthernet0	192.168.11.60	255.255.255.0

3. PC1～PC6 的地址配置与测试

参照表 7-1,对计算机进行 IP 地址的配置,配置结束后,打开 PC1 的 MS-DOS 环境,测试网络的连通性。在此,以 PC1 ping PC2 和 PC4 为例。

1) PC1 ping PC2

测试命令内容如下:

```
PC>ping 192.168.11.20

Pinging 192.168.11.20 with 32 bytes of data:
Reply from 192.168.11.20: bytes=32 time=0ms TTL=128
Reply from 192.168.11.20: bytes=32 time=0ms TTL=128
Reply from 192.168.11.20: bytes=32 time=0ms TTL=128
Reply from 192.168.11.20: bytes=32 time=0ms TTL=128
Ping statistics for 192.168.11.20:
    Packets: Sent = 4, Received = 4, Lost = 0 (0% loss),
```

```
Approximate round trip times in milli-seconds:
Minimum = 0ms, Maximum = 0ms, Average = 0ms
```

2) PC1 ping PC4

测试命令内容如下：

```
PC>ping 192.168.11.40
Pinging 192.168.11.40 with 32 bytes of data:
Reply from 192.168.11.40: bytes=32 time=0ms TTL=128
Reply from 192.168.11.40: bytes=32 time=1ms TTL=128
Reply from 192.168.11.40: bytes=32 time=0ms TTL=128
Reply from 192.168.11.40: bytes=32 time=0ms TTL=128
Ping statistics for 192.168.11.40:
    Packets: Sent = 4, Received = 4, Lost = 0 (0% loss),
Approximate round trip times in milli-seconds:
    Minimum = 0ms, Maximum = 1ms, Average = 0ms
```

从上面的测试结果可以看出，PC1与PC2、PC4的网络连通正常。事实上，交换机的所有端口默认都是开启的，且都属于VLAN 1，默认VLAN 1是关闭的。只要将计算机连接在交换机的端口上，并配置相同网段的IP地址，就可以组建成局域网，实现网络的连接畅通。

4. VLAN 的划分

1) Switch1 的配置

进入全局配置模式，配置交换机的名称，命令内容如下：

```
Switch>enable
Switch#configure terminal
Enter configuration commands, one per line. End with CNTL/Z. Switch(config)
#hostname SW1
```

(1) 在 Switch1 上创建 VLAN 10 和 VLAN 20，命令内容如下：

```
SW1(config)#vlan 10
SW1(config-vlan)#exit
SW1(config)#vlan 20
SW1(config-vlan)#exit
```

(2) 激活 VLAN 10 和 VLAN 20，命令内容如下：

```
Switch(config)#int vlan 10
SW1(config-if)#
%LINK-5-CHANGED: Interface Vlan10, changed state to up
SW1(config-if)#exit
SW1(config)#int vlan 20
```

```
SW1(config-if)#
%LINK-5-CHANGED: Interface Vlan20, changed state to up
```

(3) 将交换机 Switch1 的 f0/1 和 f0/2 接口划分到 VLAN 10 中,命令内容如下:

```
SW1(config)#interface f0/1
SW1(config-if)#switchport mode access
SW1(config-if)#switchport access vlan 10
Switch(config-if)#
%LINEPROTO-5-UPDOWN: Line protocol on Interface Vlan10, changed state to up
SW1(config)#interface f0/2
SW1(config-if)#switchport mode access
SW1(config-if)#switchport access vlan 10
```

(4) 将交换机 Switch1 的 f0/3 接口划分到 VLAN 20 中,命令内容如下:

```
SW1(config)#interface f0/3
SW1(config-if)#switchport mode access
SW1(config-if)#switchport access vlan 20
Switch(config-if)#
%LINEPROTO-5-UPDOWN: Line protocol on Interface Vlan20, changed state to up
```

(5) 查看交换机 Switch1 的 VLAN 信息,如图 7-2 所示。

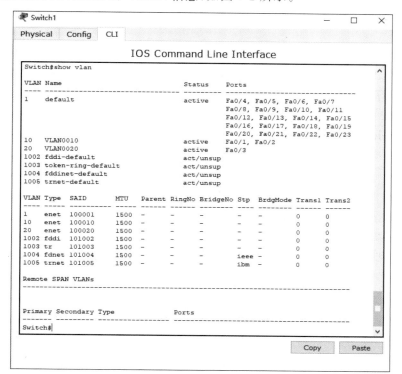

图 7-2 查看 VLAN 信息

(6) 在 Switch1 上测试 VLAN 划分后的结果。

PC1 ping PC2,命令内容如下:

```
PC>ping 192.168.11.20
Pinging 192.168.11.20 with 32 bytes of data:
Reply from 192.168.11.20: bytes=32 time=1ms TTL=128
Reply from 192.168.11.20: bytes=32 time=1ms TTL=128
Reply from 192.168.11.20: bytes=32 time=0ms TTL=128
Reply from 192.168.11.20: bytes=32 time=0ms TTL=128
Ping statistics for 192.168.11.20:
Packets: Sent = 4, Received = 4, Lost = 0 (0% loss),
Approximate round trip times in milli-seconds:
Minimum = 0ms, Maximum = 1ms, Average = 0ms
```

得到以上信息,说明 VLAN 10 中的两台计算机(PC1 和 PC2)连接畅通。

PC1 ping C4,命令内容如下:

```
PC>ping 192.168.11.40
Pinging 192.168.11.40 with 32 bytes of data:
Request timed out.
Request timed out.
Request timed out.
Request timed out.
Ping statistics for 192.168.11.40:
Packets: Sent = 4, Received = 0, Lost = 4 (100% loss)
```

以上结果说明,PC1 到 PC2 的网络畅通,PC1 到 PC4 的网络不通。这是由于 PC1 和 PC2 都属于 VLAN 10,而 PC4 属于 VLAN 20,只有相同 VLAN 中的计算机才能通信,不同 VLAN 中的计算机不能通信。

2) Switch2 的配置

按照 Switch1 的方式对 Switch2 进行配置,命令内容如下:

```
Switch>enable
Switch#configure terminal
Enter configuration commands, one per line. End with CNTL/Z.
Switch(config)#hostname SW2
SW2(config)#vlan 10
SW2(config-vlan)#exit
SW2(config)#
SW2(config)#vlan 20
SW2(config-vlan)#exit
SW2(config)#interface vlan 10
%LINK-5-CHANGED: Interface Vlan10, changed state to up
SW2(config-if)#exit
SW2(config)#interface vlan 20
```

```
%LINK-5-CHANGED: Interface Vlan20, changed state to up
SW2(config-if)#exit
SW2(config)#interface f0/1
SW2(config-if)#switchport mode access
SW2(config-if)#switchport access vlan 20
%LINEPROTO-5-UPDOWN: Line protocol on Interface Vlan20, changed state to up
SW2(config-if)#exit
SW2(config)#interface f0/2
SW2(config-if)#switchport mode access
SW2(config-if)#switchport access vlan 20
SW2(config-if)#exit
SW2(config)#interface f0/3
SW2(config-if)#switchport mode access
SW2(config-if)#switchport access vlan 10
%LINEPROTO-5-UPDOWN: Line protocol on Interface Vlan10, changed state to up
```

3）网络测试

用 PC1 ping PC3，验证跨交换机相同 VLAN 的通信，命令内容如下：

```
PC>ping 192.168.11.30
Pinging 192.168.11.30 with 32 bytes of data:
Request timed out.
Request timed out.
Request timed out.
Request timed out.
Ping statistics for 192.168.11.30:
Packets: Sent = 4, Received = 0, Lost = 4 (100% loss)
```

从上述实验可以看出，虽然两台交换机通过 f0/24 口连接，但是由于该接口没有做任何配置，属于普通接口，所以它不能传播 VLAN 信息，导致跨交换机的相同 VLAN 也不能通信。

5. trunk 链路创建

跨交换机的相同 VLAN 在通信时，需要使用 trunk。trunk 技术使得在一条物理线路上可以传送多个 VLAN 的信息，交换机从属于某一 VLAN（例如 VLAN 10）的端口接收到数据，在 trunk 链路上进行传输前，会加上一个标记，表明该数据是 VLAN 10 的；数据到了对方交换机，交换机会把该标记去掉，只发送到属于 VLAN 10 的端口上。创建 trunk 链路的命令内容如下：

```
Switch(config)#interface f0/24
Switch(config-if)#switchport mode trunk
%LINEPROTO-5-UPDOWN: Line protocol on Interface FastEthernet0/24, changed state to down
%LINEPROTO-5-UPDOWN: Line protocol on Interface FastEthernet0/24, changed state to up
```

从显示的信息中可以看到 f0/24 接口的 trunk 链路已开启。

重新测试一下 PC1 和 PC3 的网络连通性，命令内容如下：

```
PC>ping 192.168.11.30
Pinging 192.168.11.30 with 32 bytes of data:
Reply from 192.168.11.30: bytes=32 time=0ms TTL=128
Reply from 192.168.11.30: bytes=32 time=0ms TTL=128
Reply from 192.168.11.30: bytes=32 time=0ms TTL=128
Reply from 192.168.11.30: bytes=32 time=0ms TTL=128
Ping statistics for 192.168.11.30:
Packets: Sent = 4, Received = 4, Lost = 0 (0% loss),
Approximate round trip times in milli-seconds:
Minimum = 0ms, Maximum = 0ms, Average = 0ms
```

从以上信息中可以看出 PC1 和 PC3 的网络连接畅通，说明 trunk 链路实现了跨交换机相同 VLAN 内部的通信，当然也可以通过 switchport trunk allowed vlan 命令设置允许哪个 VLAN 通过。

五、实验总结

本实验可以让学生进一步认识到 VLAN 技术在网络中的作用，学会创建 VLAN，并实现基于交换机端口的 VLAN 划分，掌握 trunk 链路的作用和应用，实现 VLAN 内部的通信。

六、实验思考题

（1）VLAN 对网络广播的阻隔在第二层还是第三层？

（2）不配置 trunk 链路时，本次实验 VLAN 中的数据是否可以到达两台交换机剩余的接口？

实验八 VLAN 间路由设计

在交换机上划分 VLAN 后，VLAN 间的计算机就无法通信了。VLAN 间的通信需要借助第三层设备，可以使用路由器或三层交换机来实现这个功能，例如路由器通常会采用单臂路由模式。而在现实中，VLAN 间的路由大多是通过三层交换机实现的，三层交换机可以看成是路由器加交换机的组合，因为采用了特殊的技术，其数据处理能力较强。

VLAN 间路由

一、实验目的

(1) 熟悉路由器子接口的应用。
(2) 掌握单臂路由的原理和实现方法。
(3) 了解三层交换机的作用。
(4) 掌握基于三层交换机的 VLAN 间通信原理和应用。

二、实验环境

Cisco Packet Tracer 6.0 模拟器（二层交换机、三层交换机、路由器、计算机、直通线、Console 配置线）。

三、实验内容及步骤

(1) 搭建实验拓扑。
(2) 基于交换机端口的 VLAN 划分。
(3) 网络测试。
(4) 单臂路由应用。
(5) 基于三层交换机的 VLAN 划分及 VLAN 间通信。

四、实验过程

1. 实验拓扑

1) 单臂路由

设备选择：1841 系列路由器 1 台、2950 系列交换机 1 台、计算机 2 台。

通信线缆：直通线 4 根。

端口连接：PC1 和 PC2 分别连接在交换机的 f0/1 和 f0/2 接口上，交换机的 f0/24 连接路由器的 f0/0 接口，如图 8-1 所示。

2）三层交换

设备选择：三层交换机 1 台（本实验选择 3560 系列）、计算机 2 台。

通信线缆：直通线 2 根。

端口连接：PC1 和 PC2 分别连接在交换机的 f0/1 和 f0/2 接口上，如图 8-2 所示。

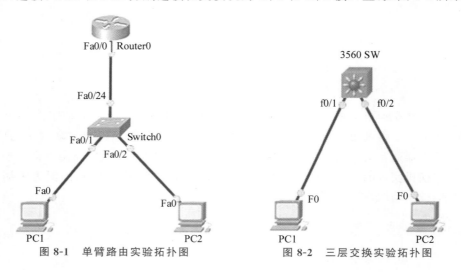

图 8-1　单臂路由实验拓扑图　　　　图 8-2　三层交换实验拓扑图

2．地址规划

既然不同的 VLAN 表示不同的网络，因此需要用不同网段的 IP 地址，具体的 IP 地址规划如表 8-1 所示。

表 8-1　IP 地址规划表

序号	设备	接口	IP 地址	子网掩码
1	PC1	FastEthernet0	192.168.10.1	255.255.255.0
2	PC2	FastEthernet0	192.168.20.2	255.255.255.0
3	Router	F0/0.1	192.168.10.100	255.255.255.0
4	Router	F0/0.2	192.168.20.100	255.255.255.0
5	3560 SW	VLAN 10	192.168.10.100	255.255.255.0
6	3560 SW	VLAN 20	192.168.20.100	255.255.255.0

3．单臂路由的实现

1）VLAN 的划分

对图 8-1 中的交换机进行 VLAN 划分，将 f0/1 接口划分到 VLAN 10 中，将 f0/2 接口划分到 VLAN 20 中，命令内容如下：

```
Switch>enable
Switch#configure terminal
    Enter configuration commands, one per line. End with CNTL/Z.
Switch (config)#vlan 10
```

```
Switch(config-vlan)#exit
Switch(config)#vlan 20
Switch(config-vlan)#exit
Switch(config)#int vlan 10
Switch(config-if)#
%LINK-5-CHANGED: Interface Vlan10, changed state to up
Switch(config-if)#exit
Switch(config)#int vlan 20
Switch(config-if)#
%LINK-5-CHANGED: Interface Vlan20, changed state to up
Switch(config)#interface range f0/1
Switch(config-range-if)#switchport mode access
Switch(config-range-if)#switchport access vlan 10
Switch(config-if)#
%LINEPROTO-5-UPDOWN: Line protocol on Interface Vlan10, changed state to up
Switch(config)#interface f0/2
Switch(config-if)#switchport mode access
Switch(config-if)#switchport access vlan 20
Switch(config-if)#
%LINEPROTO-5-UPDOWN: Line protocol on Interface Vlan20, changed state to up
```

2）VLAN 划分后的网络测试

参照表 8-1，对计算机 PC1 和 PC2 进行 IP 地址的配置。

PC1 ping PC2，命令内容如下：

```
PC>ping 192.168.20.2
Pinging 192.168.20.2 with 32 bytes of data:
Request timed out.
Request timed out.
Request timed out.
Request timed out.
Ping statistics for 192.168.20.2:
Packets: Sent = 4, Received = 0, Lost = 4 (100% loss)
```

由于在交换机上划分 VLAN 之后，不同的 VLAN 间不能通信，因此 PC1 ping PC2 不通。

3）单臂路由的配置

（1）创建 trunk 链路，命令内容如下：

```
Switch(config)#interface f0/24
Switch(config-if)#switchport mode trunk
%LINEPROTO-5-UPDOWN: Line protocol on Interface FastEthernet0/24, changed state to down
%LINEPROTO-5-UPDOWN: Line protocol on Interface FastEthernet0/24, changed state to up
```

（2）路由器的配置，命令内容如下：

```
Router>enable
Router#configure terminal
Enter configuration commands, one per line. End with CNTL/Z.
Router(config)#interface f0/0
Router(config-if)#no shutdown
Router(config-if)#exit
Router(config)#interface f0/0.1
Router(config-subif)#
%LINK-5-CHANGED: Interface FastEthernet0/0.1, changed state to up

%LINEPROTO-5-UPDOWN: Line protocol on Interface FastEthernet0/0.1, changed state to up
Router(config-subif)#encapsulation dot1Q 10
Router(config-subif)#ip address 192.168.10.100 255.255.255.0
Router(config-subif)#exit
Router(config)#interface f0/0.2
Router(config-subif)#
%LINK-5-CHANGED: Interface FastEthernet0/0.2, changed state to up

%LINEPROTO-5-UPDOWN: Line protocol on Interface FastEthernet0/0.2, changed state to up
Router(config-subif)#encapsulation dot1Q 20
Router(config-subif)#ip address 192.168.20.100 255.255.255.0
```

4）网络测试

用 PC1 ping PC2，命令内容如下：

```
PC>ping 192.168.20.2
Pinging 192.168.20.2 with 32 bytes of data:
Reply from 192.168.20.2: bytes=32 time=0ms TTL=128
Reply from 192.168.20.2: bytes=32 time=0ms TTL=128
Reply from 192.168.20.2: bytes=32 time=0ms TTL=128
Reply from 192.168.20.2: bytes=32 time=0ms TTL=128
Ping statistics for 192.168.20.2:
Packets: Sent = 4, Received = 4, Lost = 0 (0% loss),
Approximate round trip times in milli-seconds:
Minimum = 0ms, Maximum = 0ms, Average = 0ms
```

从以上信息中可以看出 PC1 和 PC2 的网络连接畅通，说明通过单臂路由实现了 VLAN 10 与 VLAN 20 间的相互通信。

4. 三层交换实现 VLAN 间通信

1）VLAN 划分

此处将划分的步骤进行稍微的变动，学生需要注意观察与之前划分 VLAN 的区别，命令内容如下：

```
Switch>enable
Switch#configure terminal
Enter configuration commands, one per line. End with CNTL/Z.
Switch(config)#interface vlan 10
Switch(config-if)#exit
Switch(config)#interface vlan 20
Switch(config-if)#exit
Switch(config)#interface f0/1
Switch(config-if)#switchport mode access
Switch(config-if)#switchport access vlan 10
% Access VLAN does not exist. Creating vlan 10
Switch(config-if)#
%LINK-5-CHANGED: Interface Vlan10, changed state to up
%LINEPROTO-5-UPDOWN: Line protocol on Interface Vlan10, changed state to up
Switch(config-if)#exit
Switch(config)#interface f0/2
Switch(config-if)#switchport mode access
Switch(config-if)#switchport access vlan 20
% Access VLAN does not exist. Creating vlan 20
Switch(config-if)#
%LINK-5-CHANGED: Interface Vlan20, changed state to up
%LINEPROTO-5-UPDOWN: Line protocol on Interface Vlan20, changed state to up
Switch(config-if)#exit
```

2）网络测试

PC1 ping PC2，命令内容如下：

```
PC>ping 192.168.20.2
Pinging 192.168.20.2 with 32 bytes of data:
Request timed out.
Request timed out.
Request timed out.
Request timed out.
Ping statistics for 192.168.20.2:
Packets: Sent = 4, Received = 0, Lost = 4 (100% loss)
```

以上信息说明 PC1 ping PC2 不通，VLAN 10 和 VLAN 20 间不能通信。

3）配置 VLAN 10 和 VLAN 20 的管理地址

具体命令内容如下：

```
Switch(config)#interface vlan 10
Switch(config-if)#ip address 192.168.10.100 255.255.255.0
Switch(config-if)#exit
```

```
Switch(config)#interface vlan 20
Switch(config-if)#ip address 192.168.20.100 255.255.255.0
Switch(config-if)#exit
```

4）开启三层交换机的路由功能

具体命令内容如下：

```
Switch(config)#ip routing
Switch(config)#
```

5）网络测试

具体命令内容如下：

```
PC>ping 192.168.20.2
Pinging 192.168.20.2 with 32 bytes of data:
Reply from 192.168.20.2: bytes=32 time=0ms TTL=128
Reply from 192.168.20.2: bytes=32 time=0ms TTL=128
Reply from 192.168.20.2: bytes=32 time=0ms TTL=128
Reply from 192.168.20.2: bytes=32 time=0ms TTL=128
Ping statistics for 192.168.20.2:
Packets: Sent = 4, Received = 4, Lost = 0 (0% loss),
Approximate round trip times in milli-seconds:
Minimum = 0ms, Maximum = 0ms, Average = 0ms
```

得到上述信息，说明在三层交换机内部实现了 VLAN 10 和 VLAN 20 间的相互通信。

五、实验总结

本实验能够让学生进一步认识到 VLAN 技术在网络中的作用，复习并掌握 VLAN 的划分方法，熟悉 trunk 链路，了解单臂路由和三层交换技术，了解路由器子接口的作用和应用，实现 VLAN 间的相互通信。

六、实验思考题

（1）路由器子接口是否要按顺序使用？

（2）你觉得单臂路由和三层交换，哪种方式的效率更高？

实验九 CDP 的配置与验证

CDP(cisco discovery protocol,Cisco 发现协议)是由 Cisco 公司推出的一种私有的二层网络协议,它能够运行在大部分的 Cisco 公司设备上。通过运行 CDP,设备能够在与它们直连的设备之间分享操作系统软件版本、IP 地址以及硬件平台等相关信息。CDP 默认每 60s 向 01-00-0C-CC-CC-CC 这个组播地址发送一次通告,如果在 180s 内未获得先前邻居设备的 CDP 通告,它将清除原来收到的 CDP 信息。因为它不依赖任何的三层协议,所以通过 CDP,可以帮助人们解决一些三层错误配置的故障,例如错误的三层地址。

CDP 的配置与验证

一、实验目的

(1) 熟悉 CDP 的作用。
(2) 掌握通过 CDP 查看邻居信息的方法。
(3) 熟悉 CDP 的配置。

二、实验环境

Cisco Packet Tracer 模拟器(三层交换机、路由器、直通线、串口模块及线缆)。

三、实验内容及步骤

(1) 学习 CDP。
(2) 搭建实验拓扑。
(3) 对设备进行配置。
(4) 分析邻居信息。

四、实验过程

1. 搭建拓扑

设备选择:3560 核心交换机 1 台、2901 路由器 2 台。
通信线缆:直通线 2 根、串口线缆 1 根。
端口连接:交换机的 F0/1 接口连接路由器 R1 的 G0/0 接口,交换机的 F0/2 接口连接路由器 R2 的 G0/0 接口,路由器 R1 与 R2 通过 S0/0/0 接口连接,如图 9-1 所示。

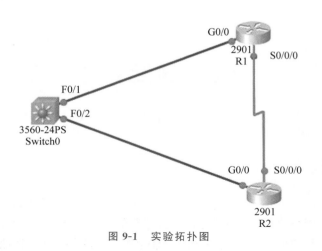

图 9-1　实验拓扑图

2. 装载模块

打开路由器 R1 的配置界面，关闭电源，装载 HWIC-2T（2 端口串行广域网接口卡），具体安装位置如图 9-2 所示。

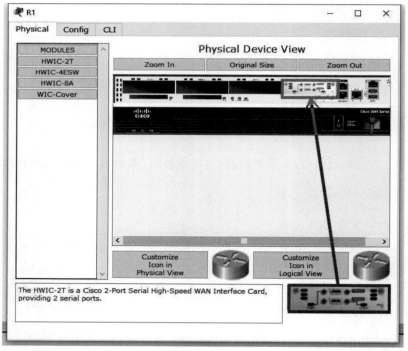

图 9-2　安装接口卡

按照同样的方式，在路由器 R2 对应的位置也安装相同的模块。

3. 设备配置

1）打开接口

具体命令内容如下：

```
R1(config)#int g0/0
R1(config-if)#no shutdown
R1(config-if)#int s0/0/0
R1(config-if)#no shutdown
R2(config)#int g0/0
R2(config-if)#no shutdown
R2(config-if)#int s0/0/0
R2(config-if)#no shutdown
R2(config-if)#clock rate 128000
```
//以上命令用于打开路由器间互连的各个接口,而默认交换机 S1 的所有接口就是打开的,因此,此
//时不做任何处理。clock rate 表示同步信号传输过程中需要协商的时钟频率,当路由器 Serial
//接口与其他路由器 Serial 接口直接对接,一边为 DTE 端,另一边为 DCE 端,clock rate 即在 DCE
//端应该设的提供数据同步时钟的命令

2) 查看 CDP 配置

具体命令内容如下:

```
R1#show cdp
Global CDP information: Global CDP information:
Sending CDP packets every 60 seconds
Sending a holdtime value of 180 seconds
Sending CDPv2 advertisements is enabled
```
//默认 CDP 是运行的,每隔 60s 从接口发送 CDP 消息,而发送出的 CDP 消息,邻居会为它保
//存 180s
```
R1#show cdp interface
Global CDP information:
Sending CDP packets every 60 seconds
Sending a holdtime value of 180 seconds
Sending CDPv2 advertisements is enabled
Router#show cdp interface
Vlan1 is administratively down, line protocol is down
Sending CDP packets every 60 seconds
Holdtime is 180 seconds
GigabitEthernet0/0 is up, line protocol is up
Sending CDP packets every 60 seconds
Holdtime is 180 seconds
GigabitEthernet0/1 is administratively down, line protocol is down
Sending CDP packets every 60 seconds
Holdtime is 180 seconds
Serial0/0/0 is up, line protocol is up
Sending CDP packets every 60 seconds
Holdtime is 180 seconds
Serial0/0/1 is administratively down, line protocol is down
Sending CDP packets every 60 seconds
```

```
Holdtime is 180 seconds
//以上内容展示了在哪些接口运行 CDP
```

3）查看 CDP 邻居

具体命令内容如下：

```
R1#show cdp neighbors
Capability Codes: R - Router, T - Trans Bridge, B - Source Route Bridge
                  S - Switch, H - Host, I - IGMP, r - Repeater, P - Phone
Device ID    Local Interface    Holdtime    Capability    Platform    Port ID
   R2         Ser 0/0/0           137         R S I        2821       Ser 0/0/0
   S1         Gig 0/0             172         S I          WS-C3560   Fas 0/1
//以上内容展示了路由器 R1 有两个邻居:R2 和 S1。Device ID 表示邻居的主机名;Local
//Interface 表明 R1 通过该接口与邻居连接,注意是 R1 上的接口;Holdtime 指收到邻居发送的 CDP
//消息的时间,采用倒计时时制;Capability 表明邻居是什么设备,第二、三行的 Capability Codes
//对各符号进行了说明;Platform 指明了邻居的硬件型号;Port ID 指明了 R1 是连接在对方哪个
//接口上的
R1#show cdp entry R2
Device ID: R2
Entry address(es):
Platform: Cisco 2821, Capabilities: Router Switch IGMP
Interface: Serial0/0/0, Port ID (outgoing port): Serial0/0/0
Holdtime : 158 sec
Version :
Cisco IOS Software, 2800 Software (C2800NM-ADVENTERPRISEK9-M), Version 12.4(11)
T1, RELEASE
SOFTWARE (fc5)
Technical Support: http://www.cisco.com/techsupport
Copyright (c) 1986-2007 by Cisco Systems, Inc.
Compiled Thu 25-Jan-07 12:50 by prod_rel_team
advertisement version: 2
VTP Management Domain: ''
//以上内容展示了邻居 R2 的详细信息,甚至可以知道邻居的 IOS 版本
R1#clear cdp table
//清除 CDP 表
```

4）关闭、开启 CDP 以及调整 CDP 参数

具体命令内容如下：

```
R1(config)#int g0/0
R1(config-if)#no cdp enable
//在 g0/0 接口上关闭 CDP,其他接口还运行 CDP
R1(config-if)#exit
R1(config)#no cdp run          //在整个路由器上关闭 CDP
```

```
R1(config)#cdp run           //在整个路由器上打开 CDP
R1(config)#cdp timer 30      //调整 CDP 消息发送时间为 30s
R1(config)#cdp holdtime 120
//调整 CDP 消息的 holdtime 为 120s,对方收到该路由器发送的 CDP 消息后将保存 120s
R1#show cdp
Global CDP information:
Sending CDP packets every 30 seconds
Sending a holdtime value of 120 seconds
Sending CDPv2 advertisements is enabled
```

五、实验总结

本实验能让学生清楚地认识到 CDP 的作用，学会搭建网络拓扑，进行 CDP 的验证，通过 CDP 查看邻居信息，熟悉接口情况。当网络出现故障时，收集设备的相关信息，能够帮助人们进行故障诊断和排除。在接触新的网络时，熟练使用 CDP，也能帮助人们熟悉设备的互连情况，更好地了解网络结构。

六、实验思考题

（1）CDP 的主要功能是什么？

（2）使用 CDP 的前提是什么？

实验十 静态路由的配置与验证

静态路由与默认路由的应用

路由器最主要的功能是数据包的转发。转发数据包时需要查找路由器的路由表信息,管理员可以通过手动的方法在路由器中直接配置路由表,这就是静态路由。虽然静态路由不适用于规模较大的网络中,但是由于静态路由简单、路由器负载小、可控性强等优点,在许多场合中还经常被使用。本实验主要介绍静态路由的配置方法,同时为以后动态路由的学习奠定基础。

一、实验目的

(1) 熟悉路由的分类。
(2) 掌握静态路由的特点。
(3) 掌握静态路由的配置方法。
(4) 熟悉静态路由的使用场合。

二、实验环境

Cisco Packet Tracer 6.0 模拟器。

三、实验内容及步骤

(1) 学习路由的基本概念。
(2) 掌握静态路由的基本知识。
(3) 搭建网络拓扑。
(4) 正确配置静态路由信息。

四、实验过程

1. 路由的基本概念

路由就是把数据从一个地方传送到另一个地方,而路由就是由路由器来完成的。路由器是一种连接多个网络或网段的网络设备,它是网络层的设备,具有判断网络地址和进行路径选择的功能。路由选择是靠路由表来实现的,路由表中有各种传输路径的数据信息。

1) 路由分类
(1) 直连:路由器自动添加和自己直接连接的网络的路由。
(2) 静态路由:管理员手动输入到路由器的路由。

(3) 动态路由：由路由协议(routing protocol)动态建立的路由。

2) 度量值

度量值(metric)是某一个路由协议判别到目的网络的最佳路径的方法。当一路由器有多条路径到达某一目的网络时，路由协议必须判断其中的哪一条是最佳的并把它放到路由表中，路由协议会为每一条路径计算出一个数值，这个数值就是度量值，通常这个值是没有单位的。度量值越小，路由越优先。然而不同的路由协议定义度量值的方法是不一样的，所以不同的路由协议选择出的最佳距离可能也不一样。

2. 静态路由

(1) 概念：静态路由是由管理员在路由器中手动配置的固定路由，路由明确地指定了包到达目的地必须经过的路径，静态路由不会随网络的变化而变化，所以，静态路由一般用于网络规模不大、拓扑结构相对固定的网络。

(2) 优点：对路由的行为进行精确的控制；减少了网络流量，不会占用路由器太多的CPU 和 RAM 资源，也不占用线路的带宽；是单向的；配置简单。

(3) 缺点：不能动态反映网络拓扑，当网络拓扑发生变化时，管理员就必须手动改变路由表。

3. 实验拓扑

通过学习静态路由知识，结合专业培养和实验目的，设计网络拓扑结构，如图 10-1 所示。

设备选择：1841 系列路由器 2 台、2811 系列路由器 1 台、计算机 2 台。

通信线缆：交叉线 4 根。

端口连接：PC1 连接 Router1 的 f0/1 接口，Router1 的 f0/0 接口连接 Router2 的 f0/1 接口，Router2 的 f0/0 接口连接 Router3 的 f0/1 接口，PC2 连接 Router3 的 f0/0 接口。

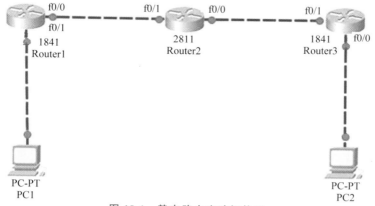

图 10-1 静态路由实验拓扑图

4. 静态路由配置

1) IP 地址规划

IP 地址规划如表 10-1 所示。

表 10-1 IP 地址规划表

序 号	设 备	接 口	IP 地址	子网掩码
1	PC1	FastEthernet0	192.168.30.10	255.255.255.0
2	Router1	FastEthernet 0/1	192.168.30.1	255.255.255.0
3	Router1	FastEthernet 0/0	192.168.31.1	255.255.255.0
4	Router2	FastEthernet 0/1	192.168.31.2	255.255.255.0
5	Router2	FastEthernet 0/0	192.168.32.1	255.255.255.0
6	Router3	FastEthernet 0/1	192.168.32.2	255.255.255.0
7	Router3	FastEthernet 0/0	192.168.33.1	255.255.255.0
8	PC2	FastEthernet0	192.168.33.20	255.255.255.0

2）设备基本配置

（1）PC1 的配置参照表 10-1 的信息，配置 PC1 的 IP 地址如图 10-2 所示。

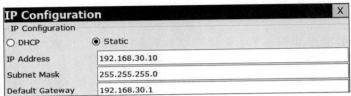

图 10-2　PC1 的 IP 地址配置

（2）Router1 的配置，命令内容如下：

```
Router>enable
Router#configure terminal
Router(config)#hostname R1
R1(config)#interface FastEthernet 0/1
R1(config-if)#no shutdown
R1(config-if)#ip address 192.168.30.1 255.255.255.0
R1(config-if)#exit
R1(config)#interface FastEthernet 0/0
R1(config-if)#no shutdown
R1(config-if)#ip address 192.168.31.1 255.255.255.0
```

（3）Router2 的配置，命令内容如下：

```
Router>enable
Router#configure terminal
Router(config)#hostname R2
R2(config)#interface FastEthernet 0/1
R2(config-if)#no shutdown
R2(config-if)#ip address 192.168.31.2 255.255.255.0
```

```
R2(config-if)#exit
R2(config)#interface FastEthernet 0/0
R2(config-if)#no shutdown
R2(config-if)#ip address 192.168.32.1 255.255.255.0
```

(4) Router3 的配置,命令内容如下:

```
Router>enable
Router#configure terminal
Router(config)#hostname R3
R3(config)#interface FastEthernet 0/1
R3(config-if)#no shutdown
R3(config-if)#ip address 192.168.32.2 255.255.255.0
R3(config-if)#exit
R3(config)#interface FastEthernet 0/0
R3(config-if)#no shutdown
R3(config-if)#ip address 192.168.33.1 255.255.255.0
```

(5) PC2 的配置参照表 10-1 的信息,配置 PC2 的 IP 地址如图 10-3 所示。

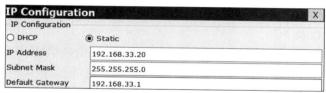

图 10-3　PC2 的 IP 地址配置

3) 网络测试

配置结束后,会发现拓扑中所有的结点都由原来的红灯变为绿灯。接下来做网络测试,打开 PC1 的 MS-DOS 环境,测试它到其他网络的连通性。

(1) 测试网关,命令内容如下:

```
PC>ping 192.168.30.1
Pinging 192.168.30.1 with 32 bytes of data:
Reply from 192.168.30.1: bytes=32 time=1ms TTL=255
Reply from 192.168.30.1: bytes=32 time=0ms TTL=255
Ping statistics for 192.168.30.1:
Packets: Sent = 2, Received = 2, Lost = 0 (0% loss),
Approximate round trip times in milli-seconds:
Minimum = 0ms, Maximum = 1ms, Average = 0ms
```

得到上述信息,就说明 PC1 到网关是畅通的。

(2) 测试 PC1 到直连路由器的 f0/0 接口,命令内容如下:

```
PC>ping 192.168.31.1
Pinging 192.168.31.1 with 32 bytes of data:
```

```
Reply from 192.168.31.1: bytes=32 time=0ms TTL=255
Reply from 192.168.31.1: bytes=32 time=0ms TTL=255
Ping statistics for 192.168.31.1:
Packets: Sent = 2, Received = 2, Lost = 0 (0% loss),
Approximate round trip times in milli-seconds:
Minimum = 0ms, Maximum = 0ms, Average = 0ms
```

得到以上信息,说明 PC1 到直连路由器的 f0/0 接口是畅通的。

(3) 测试 PC1 与邻居路由器 Router2 的 f0/1 接口是否可达,命令内容如下:

```
PC>ping 192.168.31.2
Pinging 192.168.31.2 with 32 bytes of data:
Request timed out.
Request timed out.
Ping statistics for 192.168.31.2:
Packets: Sent = 2, Received = 0, Lost = 2 (100% loss)
```

得到以上信息,说明 PC1 与 Router2 的 f0/1 接口不通,因此很显然,信息要到达 Router3 与 PC2 就更不可能了。

按照同样的方式,也可以从 PC2 开始做测试,同样会发现,信息无法到达 Router2 的 f0/0 接口、Router1 以及 PC1。

造成上述网络故障的原因就是路由器中没有对应的路由表信息,从而导致信息不可达,接下来配置静态路由信息。

4) 静态路由的配置

格式为:ip route+目的网络+目的网络子网掩码+下一跳地址。下一跳地址实际上就是与邻居路由器的直连接口地址。

(1) Router1 的配置,命令内容如下:

```
R1(config)#ip route 192.168.32.0 255.255.255.0 192.168.31.2
R1(config)#ip route 192.168.33.0 255.255.255.0 192.168.31.2
```

(2) Router2 的配置,命令内容如下:

```
R2(config)#ip route 192.168.30.0 255.255.255.0 192.168.31.1
R2(config)#ip route 192.168.33.0 255.255.255.0 192.168.32.2
```

(3) Router3 的配置,命令内容如下:

```
R3(config)#ip route 192.168.31.0 255.255.255.0 192.168.32.1
R3(config)#ip route 192.168.30.0 255.255.255.0 192.168.32.1
```

5) 网络测试

经过前文的网络测试,可以看出,配置静态路由之前,数据无法从 PC1 到达 Router2 的 f0/1 接口、Router3 以及 PC2。现在,进行重新测试。

（1）用 PC1 ping Router2 的 f0/1 接口,命令内容如下:

```
PC>ping 192.168.31.2
Pinging 192.168.31.2 with 32 bytes of data:
Reply from 192.168.31.2: bytes=32 time=1ms TTL=254
Reply from 192.168.31.2: bytes=32 time=0ms TTL=254
Ping statistics for 192.168.31.2:
Packets: Sent = 2, Received = 2, Lost = 0 (0% loss),
Approximate round trip times in milli-seconds:
Minimum = 0ms, Maximum = 1ms, Average = 0ms
```

得到以上信息,说明从 PC1 到达 192.168.31.0 的网络畅通。

（2）用 PC1 ping Router3 的 f0/1 接口,命令内容如下:

```
PC>ping 192.168.32.2
Pinging 192.168.32.2 with 32 bytes of data:
Reply from 192.168.32.2: bytes=32 time=1ms TTL=253
Reply from 192.168.32.2: bytes=32 time=13ms TTL=253
Ping statistics for 192.168.32.2:
Packets: Sent = 2, Received = 2, Lost = 0 (0% loss),
Approximate round trip times in milli-seconds:
Minimum = 1ms, Maximum = 13ms, Average = 7ms
```

得到以上信息,说明从 PC1 到达 192.168.32.0 的网络畅通。

（3）用 PC1 ping PC2,命令内容如下:

```
PC>ping 192.168.33.20
Pinging 192.168.33.20 with 32 bytes of data:
Reply from 192.168.33.20: bytes=32 time=0ms TTL=125
Reply from 192.168.33.20: bytes=32 time=19ms TTL=125
Ping statistics for 192.168.33.20:
Packets: Sent = 2, Received = 2, Lost = 0 (0% loss),
Approximate round trip times in milli-seconds:
Minimum = 0ms, Maximum = 19ms, Average = 9ms
```

得到以上信息,说明从 PC1 到达 192.168.33.0 的网络畅通。

五、实验总结

本实验让学生能够清楚地认识路由表的作用,学会搭建网络拓扑,对路由器进行基本配置,熟练掌握静态路由的概念和配置方法。当网络出现故障时,要学会按照上述的方法,一步一步检测连通性,找出网络故障的位置,认真分析,发现网络故障的原因,能够通过在静态路由的应用解决网络故障。

六、实验思考题

（1）实验中下一跳地址用的是 IP 地址,能否换成接口(如 f0/0)?
（2）如果再增加一台路由器,还需要配置多少条路由?

实验十一 默认路由的配置与验证

默认路由是一种特殊的静态路由,指的是当路由表中与包的目的地址之间没有匹配的表项时路由器能够做出的选择。如果没有默认路由,那么目的地址在路由表中将因没有匹配表项的包而被丢弃,默认路由在某些时候非常有效,当存在末梢网络时,默认路由会大大简化路由器的配置,减轻管理员的工作负担,提高网络性能。

一、实验目的

(1) 熟悉默认路由的概念。
(2) 熟悉默认路由的特点。
(3) 掌握默认路由的应用及配置方法。

二、实验环境

Cisco Packet Tracer 6.0 模拟器。

三、实验内容及步骤

(1) 了解默认路由的概念。
(2) 掌握默认路由的特点。
(3) 搭建网络拓扑。
(4) 正确配置默认路由信息。

四、实验过程

1. 了解默认路由的基本概念

默认路由(default route)是当 IP 数据包中的目的地址找不到存在的其他路由时,路由器所选择的路由。目的地址不在路由器路由表里的所有数据包都会使用默认路由,这条路由一般会通往另一台路由器,而这台路由器也同样会处理数据包。如果知道应该怎么路由这个数据包,则数据包会被转发到已知的路由;否则,数据包会被转发到默认路由,从而到达另一台路由器。每次转发,路由都增加了一跳的距离。

2. 默认路由的特点

(1) 配置简单,易于实现。

（2）减少了路由的条目。

（3）使用默认路由时，不用熟悉外部网络的具体结构。

（4）优先级最低，只有当路由器没有发现匹配报文目的 IP 地址的任何具体路由之后，才会使用默认路由来转发数据。

3. 实验拓扑

通过学习静态路由知识，结合专业培养和实验目的，设计网络拓扑结构，如图 11-1 所示。

设备选择：1841 系列路由器 2 台、2811 系列路由器 1 台、计算机 2 台。

通信线缆：交叉线 4 根。

端口连接：PC1 连接 Router1 的 f0/1 接口，Router1 的 f0/0 接口连接 Router2 的 f0/1 接口，Router2 的 f0/0 接口连接 Router3 的 f0/1 接口，PC2 连接 Router3 的 f0/0 接口。

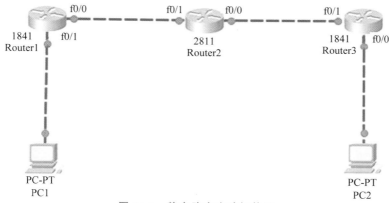

图 11-1 静态路由实验拓扑图

4. 默认路由的配置

本实验，为了验证默认路由，首先，按照静态路由的配置方式，完成前期的基本配置，并配置路由器 Router1 和 Router2 的静态路由信息；其次，测试网络的连通性；最后，在 Router3 上配置默认路由，并测试配置后的结果，验证默认路由。

1）IP 地址规划

IP 地址规划如表 11-1 所示。

表 11-1 IP 地址规划表

序号	设备	接口	IP 地址	子网掩码
1	PC1	FastEthernet0	192.168.30.10	255.255.255.0
2	Router1	FastEthernet 0/1	192.168.30.1	255.255.255.0
3	Router1	FastEthernet 0/0	192.168.31.1	255.255.255.0
4	Router2	FastEthernet 0/1	192.168.31.2	255.255.255.0
5	Router2	FastEthernet 0/0	192.168.32.1	255.255.255.0

续表

序 号	设 备	接 口	IP 地址	子网掩码
6	Router3	FastEthernet 0/1	192.168.32.2	255.255.255.0
7	Router3	FastEthernet 0/0	192.168.33.1	255.255.255.0
8	PC2	FastEthernet0	192.168.33.20	255.255.255.0

2) 设备基本配置

（1）PC1 的配置参照表 11-1 的信息，配置 PC1 的 IP 地址如图 11-2 所示。

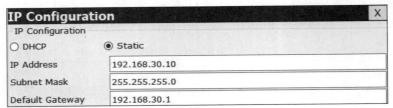

图 11-2　PC1 的 IP 地址配置

（2）Router1 的配置，命令内容如下：

```
Router>enable
Router#configure terminal
Router(config)#hostname R1
R1(config)#interface FastEthernet 0/1
R1(config-if)#no shutdown
R1(config-if)#ip address 192.168.30.1 255.255.255.0
R1(config-if)#exit
R1(config)#interface FastEthernet 0/0
R1(config-if)#no shutdown
R1(config-if)#ip address 192.168.31.1 255.255.255.0
```

（3）Router2 的配置，命令内容如下：

```
Router>enable
Router#configure terminal
Router(config)#hostname R2
R2(config)#interface FastEthernet 0/1
R2(config-if)#no shutdown
R2(config-if)#ip address 192.168.31.2 255.255.255.0
R2(config-if)#exit
R2(config)#interface FastEthernet 0/0
R2(config-if)#no shutdown
R2(config-if)#ip address 192.168.32.1 255.255.255.0
```

（4）Router3 的配置，命令内容如下：

```
Router>enable
Router#configure terminal
Router(config)#hostname R3
R3(config)#interface FastEthernet 0/1
R3(config-if)#no shutdown
R3(config-if)#ip address 192.168.32.2 255.255.255.0
R3(config-if)#exit
R3(config)#interface FastEthernet 0/0
R3(config-if)#no shutdown
R3(config-if)#ip address 192.168.33.1 255.255.255.0
```

（5）PC2 的配置参照表 11-1 的信息，配置 PC2 的 IP 地址如图 11-3 所示。

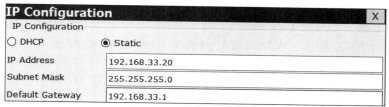

图 11-3　PC2 的 IP 地址配置

3）网络测试

配置结束后，会发现拓扑中所有的结点都由原来的红灯变为绿灯，接下来做网络测试，打开 PC1 的 MS-DOS 环境，测试它到其他网络的连通性。

（1）测试网关，命令内容如下：

```
PC>ping 192.168.30.1
Pinging 192.168.30.1 with 32 bytes of data:
Reply from 192.168.30.1: bytes=32 time=1ms TTL=255
Reply from 192.168.30.1: bytes=32 time=0ms TTL=255
Ping statistics for 192.168.30.1:
Packets: Sent = 2, Received = 2, Lost = 0 (0% loss),
Approximate round trip times in milli-seconds:
Minimum = 0ms, Maximum = 1ms, Average = 0ms
```

得到上述信息，说明 PC1 到网关是连通的。

（2）测试 PC1 到直连路由器的 f0/0 接口，命令内容如下：

```
PC>ping 192.168.31.1
Pinging 192.168.31.1 with 32 bytes of data:
Reply from 192.168.31.1: bytes=32 time=0ms TTL=255
Reply from 192.168.31.1: bytes=32 time=0ms TTL=255
Ping statistics for 192.168.31.1:
```

```
Packets: Sent = 2, Received = 2, Lost = 0 (0% loss),
Approximate round trip times in milli-seconds:
Minimum = 0ms, Maximum = 0ms, Average = 0ms
```

得到以上信息,说明 PC1 到直连路由器的 f0/0 接口是连通的。
(3)测试 PC1 到邻居路由器 Router2 的 f0/1 接口是否可达,命令内容如下:

```
PC>ping 192.168.31.2
Pinging 192.168.31.2 with 32 bytes of data:
Request timed out.
Request timed out.
Ping statistics for 192.168.31.2:
Packets: Sent = 2, Received = 0, Lost = 2 (100% loss)
```

得到以上信息,说明 PC1 与 Router2 的 f0/1 接口不通,因此很显然,信息要到达 Router3 与 PC2 就更不可能了。

按照同样的方式,也可以从 PC2 开始做测试,同样会发现,信息无法到达 Router2 的 f0/0 接口、Router1 以及 PC1。

由于路由器中没有对应的路由表信息,导致信息不可达,接下来配置路由信息。

4)在路由器 Router1 和 Router2 上的静态路由配置
(1)Router1 的配置,命令内容如下:

```
R1(config)#ip route 192.168.32.0 255.255.255.0 192.168.31.2
R1(config)#ip route 192.168.33.0 255.255.255.0 192.168.31.2
```

(2)Router2 的配置,命令内容如下:

```
R2(config)#ip route 192.168.30.0 255.255.255.0 192.168.31.1
R2(config)#ip route 192.168.33.0 255.255.255.0 192.168.32.2
```

5)网络测试
ping 192.168.31.1 的命令内容如下:

```
R3#ping 192.168.31.1
Type escape sequence to abort.
Sending 5, 100-byte ICMP Echos to 192.168.31.1, timeout is 2 seconds:
.....
Success rate is 0 percent (0/5)
```

出现"."说明到达 192.168.31.1 的网络不通。
ping 192.168.30.1 的命令内容如下:

```
R3#ping 192.168.30.1
Type escape sequence to abort.
```

```
Sending 5, 100-byte ICMP Echos to 192.168.30.1, timeout is 2 seconds:
.....
Success rate is 0 percent (0/5)
```

出现"."说明到达 192.168.30.1 的网络不通。

从上面的测试可以看出,Router3 到 192.168.31.0 和 192.168.30.0 的网络都是不通的,其原因就是没有到达这两个网段的路由信息,可以通过 show ip route 命令来查看路由器 Router3 上的路由条目。

6) 配置默认路由

接下来在路由器 Router3 上配置默认路由,命令如下:

```
R3(config)#ip route 0.0.0.0 0.0.0.0 192.168.32.1
```

该命令的格式与配置静态路由没有区别,也可以说默认路由就是一种特殊的静态路由,也包括目的网络号、目的网络子网掩码以及下一跳地址,此时全为 0 的网络号和子网掩码表示任何网络,整个路由信息的含义就是:到达任何一个网络都只需要把信息传递给192.168.32.1。

7) 网络测试

具体命令内容如下:

```
R3#ping 192.168.31.1
Type escape sequence to abort.
Sending 5, 100-byte ICMP Echos to 192.168.31.1, timeout is 2 seconds:
..!!!
Success rate is 60 percent (3/5), round-trip min/avg/max = 0/0/1 ms
R3#ping 192.168.30.1
Type escape sequence to abort.
Sending 5, 100-byte ICMP Echos to 192.168.30.1, timeout is 2 seconds:
!!!!!
Success rate is 100 percent (5/5), round-trip min/avg/max = 0/0/0 ms
```

出现"!"说明网络连通。

五、实验总结

本实验可以让学生进一步认识路由表的作用,学会搭建网络拓扑,对路由器进行基本配置,熟练掌握默认路由的概念和配置方法,当网络出现故障时,能学会按照上述方法,一步一步检测连通性,找出网络故障的位置,认真分析,得出网络故障的原因,能够通过默认路由的应用解决网络故障。

六、实验思考题

(1) 本实验中,路由器 Router2 上若配置默认路由,全网是否可通?
(2) 默认路由的使用规则是什么?

实验十二 RIP 的配置与验证

RIP 的应用

动态路由协议包括距离向量路由协议和链路状态路由协议,而 RIP(routing information protocols,路由信息协议)就是使用最广泛的距离向量路由协议。RIP 是为小型网络环境设计的,因为这类协议的路由学习及路由更新将产生较大的流量,占用过多的带宽。采用 RIP 的路由器不知道网络的全局情况,如果路由更新在网络上传播慢,将会导致网络收敛较慢,造成路由环路,RIP 采用水平分割、毒性逆转、定义最大跳数、闪式更新、抑制计时 5 个机制来避免路由环路。

RIP 协议的特征:①是距离向量路由协议;②使用跳数(hop count)作为度量值;③默认路由更新周期为 30s;④管理距离(AD)为 120;⑤支持触发更新;⑥最大跳数为 15 跳;⑦支持等价路径,默认为 4 条,最多 6 条;⑧使用 UDP 520 端口进行路由更新。

RIPv1:在路由更新的过程中不携带子网信息;不提供认证;不支持 VLSM 和 CIDR;采用广播更新;是有类别(classful)路由协议。

RIPv2:在路由更新的过程中携带子网信息;提供明文和 MD5 认证;支持 VLSM 和 CIDR;采用组播(224.0.0.9)更新;是无类别(classless)路由协议。

一、实验目的

(1) 在路由器上启动 RIPv2 路由进程。
(2) 掌握 RIPv2 的配置方法。
(3) 了解路由协议的自动汇总。
(4) 查看和调试 RIPv2 相关信息。

二、实验环境

Cisco Packet Tracer 6.0 模拟器。

三、实验内容及步骤

(1) 搭建 RIP 的实验拓扑。
(2) 进行网络地址规划。
(3) 配置路由器端口。
(4) RIP 的配置和应用。
(5) 查看和测试 RIPv2 信息。

四、实验过程

1. 实验拓扑

通过学习 RIP 的知识,结合专业培养和实验要求,设计网络拓扑结构,如图 12-1 所示。

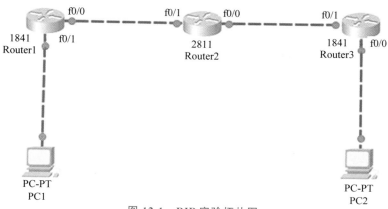

图 12-1 RIP 实验拓扑图

设备选择:1841 系列路由器 2 台、2811 系列路由器 1 台、计算机 2 台。

通信线缆:交叉线 4 根。

端口连接:PC1 连接 Router1 的 f0/1 接口,Router1 的 f0/0 接口连接 Router2 的 f0/1 接口,Router2 的 f0/0 接口连接 Router3 的 f0/1 接口,PC2 连接 Router3 的 f0/0 接口。

2. 地址规划

IP 地址规划如表 12-1 所示。

表 12-1 IP 地址规划表

序号	设备	接口	IP 地址	子网掩码
1	PC1	FastEthernet0	192.168.11.10	255.255.255.0
2	PC2	FastEthernet0	192.168.32.20	255.255.255.0
3	Router1	FastEthernet 0/1	192.168.11.1	255.255.255.0
4	Router1	FastEthernet 0/0	192.168.12.1	255.255.255.0
5	Router2	FastEthernet 0/1	192.168.12.2	255.255.255.0
6	Router2	FastEthernet 0/0	192.168.23.2	255.255.255.0
7	Router3	FastEthernet 0/1	192.168.23.3	255.255.255.0
8	Router3	FastEthernet 0/0	192.168.32.3	255.255.255.0

3. 路由器接口基本配置

(1) Router1 的配置,命令内容如下:

```
Router>enable
Router#configure terminal
```

```
Router(config)#hostname R1
R1(config)#interface FastEthernet 0/1
R1(config-if)#ip address 192.168.11.1 255.255.255.0
R1(config-if)#no shutdown
R1(config-if)#exit
R1(config)#interface FastEthernet 0/0
R1(config-if)#ip address 192.168.12.1 255.255.255.0
R1(config-if)#no shutdown
R1(config-if)#exit
```

（2）Router2 的配置，命令内容如下：

```
Router>enable
Router#configure terminal
Router(config)#hostname R2
R2(config)#interface FastEthernet 0/1
R2(config-if)#ip address 192.168.12.2 255.255.255.0
R2(config-if)#no shutdown
R2(config-if)#exit
R2(config)#interface FastEthernet 0/0
R2(config-if)#ip address 192.168.23.2 255.255.255.0
R2(config-if)#no shutdown
```

（3）Router3 的配置，命令内容如下：

```
Router>enable
Router#configure terminal
Router(config)#hostname R3
R3(config)#interface FastEthernet 0/1
R3(config-if)#ip address 192.168.23.3 255.255.255.0
R3(config-if)#no shutdown
R3(config-if)#exit
R3(config)#interface FastEthernet 0/0
R3(config-if)#ip address 192.168.32.3 255.255.255.0
R3(config-if)#no shutdown
```

4. PC1 和 PC2 的配置

配置 PC1 和 PC2 时要注意它们所在的网段以及网关地址，网关地址为它们直连路由器接口的 IP 地址，如图 12-2 和图 12-3 所示。

5. 网络测试

ping Router1 自身接口，如图 12-4 所示。

得到以上信息，说明 Router1 的 f0/0 和 f0/1 接口配置正确，正常运行。

ping Router1 直连的设备接口，如图 12-5 所示。

得到以上信息，说明 Router1 到 PC1 和 Router1 的 f0/1 接口数据可达，即 Router1 到

实验十二　RIP 的配置与验证

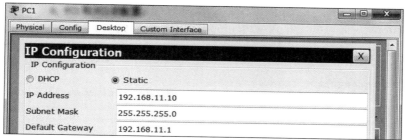

图 12-2　PC1 的配置信息

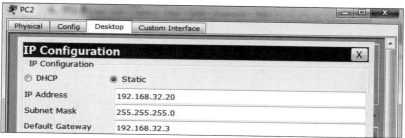

图 12-3　PC2 的配置信息

```
R1#ping 192.168.11.1

Type escape sequence to abort.
Sending 5, 100-byte ICMP Echos to 192.168.11.1, timeout is 2 seconds:
!!!!!
Success rate is 100 percent (5/5), round-trip min/avg/max = 1/6/20 ms

R1#ping 192.168.12.1

Type escape sequence to abort.
Sending 5, 100-byte ICMP Echos to 192.168.12.1, timeout is 2 seconds:
!!!!!
Success rate is 100 percent (5/5), round-trip min/avg/max = 1/8/24 ms
```

图 12-4　ping Router1 自身接口

```
R1#ping 192.168.11.10

Type escape sequence to abort.
Sending 5, 100-byte ICMP Echos to 192.168.11.10, timeout is 2 seconds:
!!!!!
Success rate is 100 percent (5/5), round-trip min/avg/max = 0/0/1 ms

R1#ping 192.168.12.2

Type escape sequence to abort.
Sending 5, 100-byte ICMP Echos to 192.168.12.2, timeout is 2 seconds:
!!!!!
Success rate is 100 percent (5/5), round-trip min/avg/max = 0/0/1 ms
```

图 12-5　ping Router1 直连的设备接口

达 192.168.11.0 和 192.168.12.0 网络连通，其实 192.168.11.0 和 192.168.12.0 就是 Router1 直连的两个网络。

ping Router1 到 Router2 的 f0/0 接口，如图 12-6 所示。

```
R1#ping 192.168.23.2

Type escape sequence to abort.
Sending 5, 100-byte ICMP Echos to 192.168.23.2, timeout is 2 seconds:
.....
Success rate is 0 percent (0/5)
```

图 12-6　ping Router1 到 Router2 的 f0/0 接口

得到以上信息，说明 Router1 到 Router2 的 f0/0 接口不通，因为 192.168.23.0 不是 Router1 的直连网络，Router1 上没有到达 192.168.23.0 网络的路由。

6. 查看路由信息

Router1 的路由表信息如图 12-7 所示。

```
R1#show ip route
Codes: C - connected, S - static, I - IGRP, R - RIP, M - mobile, B - BGP
       D - EIGRP, EX - EIGRP external, O - OSPF, IA - OSPF inter area
       N1 - OSPF NSSA external type 1, N2 - OSPF NSSA external type 2
       E1 - OSPF external type 1, E2 - OSPF external type 2, E - EGP
       i - IS-IS, L1 - IS-IS level-1, L2 - IS-IS level-2, ia - IS-IS
inter area
       * - candidate default, U - per-user static route, o - ODR
       P - periodic downloaded static route

Gateway of last resort is not set

C    192.168.11.0/24 is directly connected, FastEthernet0/1
C    192.168.12.0/24 is directly connected, FastEthernet0/0
```

图 12-7　查看 Router1 的路由表信息

图 12-7 中 C 表示与 Router1 直连的网，可以看到只有两条直连网络 192.168.11.0 和 192.168.12.0，也就说明除了这两条，Router1 到其他网络都是不通的。

7. 配置动态 RIP 路由信息

（1）Router1 的配置，命令内容如下：

```
R1(config)#router rip
R1(config-router)#version 2
R1(config-router)#network 192.168.11.0
R1(config-router)#network 192.168.12.0
```

（2）Router2 的配置，命令内容如下：

```
R2(config)#router rip
R2(config-router)#version 2
R2(config-router)#network 192.168.12.0
R2(config-router)#network 192.168.23.0
```

(3) Router3 的配置，命令内容如下：

```
R3(config)#router rip
R3(config-router)#version 2
R3(config-router)#network 192.168.23.0
R3(config-router)#network 192.168.32.0
```

8. 查看路由信息

在 Router1 的特权模式下，运行命令 show ip route 即可查看路由信息，具体内容如下：

```
R1#show ip route
Codes: C-connected, S-static, I-IGRP, R-RIP, M-mobile, B-BGP
D - EIGRP, EX - EIGRP external, O - OSPF, IA - OSPF inter area
N1 - OSPF NSSA external type 1, N2 - OSPF NSSA external type 2
E1 - OSPF external type 1, E2 - OSPF external type 2, E - EGP
i - IS-IS, L1-IS-IS level-1, L2-IS-IS level-2, ia - IS-IS inter area
* - candidate default, U - per-user static route, o - ODR
P - periodic downloaded static route
Gateway of last resort is not set
C 192.168.11.0/24 is directly connected, FastEthernet0/1
C 192.168.12.0/24 is directly connected, FastEthernet0/0
R 192.168.23.0/24[120/1] via 192.168.12.2, 00:00:09, FastEthernet0/0
R 192.168.32.0/24[120/2] via 192.168.12.2, 00:00:05, FastEthernet0/0
```

从以上信息能够清楚地看到，在 Router1 上除了两条直连的网络外，还多了到达 192.168.23.0 和 192.168.32.0 网络的路由。其中，R 表示是通过 RIP 学习到的。

按照同样的方法，可以自己尝试查看 Router2 和 Router3 上的路由信息。

9. 网络测试

以 PC1 ping PC2 为例，验证一下网络的连通性，命令内容如下：

```
PC>ping 192.168.32.20

Pinging 192.168.32.20 with 32 bytes of data:
Reply from 192.168.32.20: bytes=32 time=1ms TTL=125
Reply from 192.168.32.20: bytes=32 time=11ms TTL=125
Reply from 192.168.32.20: bytes=32 time=10ms TTL=125
Reply from 192.168.32.20: bytes=32 time=1ms TTL=125
Ping statistics for 192.168.32.20:
Packets: Sent = 4, Received = 4, Lost = 0 (0% loss),
Approximate round trip times in milli-seconds:
Minimum = 1ms, Maximum = 11ms, Average = 5ms
```

得到以上信息说明网络连通。

五、实验总结

本次实验能够让学生进一步认识到路由表的作用，学会创建网络拓扑，巩固对路由器的基本配置，熟悉 RIP 的概念和配置方法，掌握常用命令的应用。当网络出现故障时，要学会按照上述的方法，一步一步检测连通性，找出网络故障的位置，认真分析，得出网络故障的原因，能够通过静态路由的应用解决网络故障。

  这里对 RIP 常用命令进行汇总，具体如下。

  show ip route：查看路由表。

  show ip protocols：查看 IP 路由协议配置和统计信息。

  show ip rip database：查看 RIP 数据库。

  debug ip rip：动态查看 RIP 的更新过程。

  clear ip route *：清除路由表。

  router rip：启动 RIP 进程。

  network：通告网络。

  version：定义 RIP 的版本。

  no auto-summary：关闭自动汇总。

  ip rip send version：配置 RIP 发送的版本。

  ip rip receive version：配置 RIP 接收的版本。

  passive-interface：配置被动接口。

  neighbor：配置单播更新的目标。

  ip summary-address rip：配置 RIP 手动汇总。

  key key-id：配置 Key ID。

  key-string：配置 Key ID 的密钥。

  ip rip triggered：配置触发更新。

  ip rip authentication mode：配置认证模式。

  ip rip authentication key-chain：配置认证使用的钥匙链。

  timers basic：配置更新的计时器。

  maximum-paths：配置等价路径的最大值。

  ip default-network：向网络中注入默认路由。

六、实验思考题

  (1) 当网络的跳数达到 16 时，采用 RIP，是否可以连通？

  (2) 网络结构发生变化，RIP 能否立即更新？

实验十三 网络应用服务平台的构建与测试

网络服务

一、实验目的

（1）熟悉网络应用服务的概念。
（2）掌握 DHCP 和 WWW 服务器的构建和应用。
（3）掌握 DNS 和 FTP 服务器的构建及应用。

二、实验环境

Cisco Packet Tracer 6.0 模拟器（路由器、交换机、服务器、计算机、交叉线以及直通线）。

三、实验内容及步骤

（1）DHCP 服务器的配置与应用。
（2）WWW 服务器的配置与应用。
（3）FTP 服务器的配置与应用。
（4）DNS 服务器的配置与应用。

四、实验过程

1. 实验拓扑

设备选择：1841 系列路由器 1 台、2950 系列交换机 1 台、服务器 3 台、计算机 3 台。

通信线缆：直通线 5 根、交叉线 2 根。

端口连接：PC1、PC2 和 PC3 分别连接在交换机的 f0/1、f0/2 和 f0/3 接口上，交换机的 f0/24 接口连接在路由器的 g0/2 接口上，DHCP 服务器连接在交换机的 f0/10 接口，FTP 服务器连接在路由器的 g0/1 接口上，WWW 服务器连接在路由器的 g0/0 接口上，如图 13-1 所示。

2. DHCP 服务器的搭建

1）DHCP 服务器的地址规划

DHCP 服务器本身就是网络中的一台主机，它必须有一个确定而有效的 IP 地址，这样才能访问到它，DHCP 服务器的地址规划如图 13-2 所示。

2）DHCP 服务器的配置

构建 DHCP 服务器时，首先需要开启 DHCP 服务，然后设置名称，配置默认

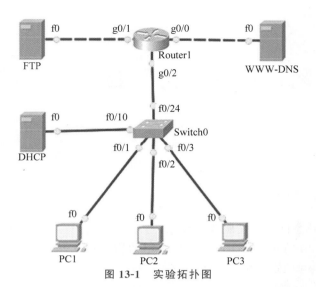

图 13-1 实验拓扑图

图 13-2 DHCP 服务器的地址规划

网关以及 DNS 域名解析服务器地址,接下来比较重要的部分就是配置地址池,需要设置起始地址、子网掩码以及可提供的地址数,最后单击 Add 按钮进行添加。DHCP 服务器的配置如图 13-3 所示。

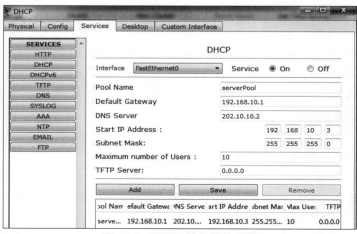

图 13-3 DHCP 服务器的配置

在一个 DHCP 服务器上可以建立多个地址池,如果需要修改哪个地址池,可以选中后直接修改,修改完成后单击 Save 按钮即可。

3) 动态地址分配

打开计算机的 IP 信息配置界面,选择 DHCP 单选按钮,计算机就会向网络中的 DHCP 服务器发送获取地址的请求,随后显示请求结果,PC1、PC2 和 PC3 的地址分配结果如图 13-4～图 13-6 所示。

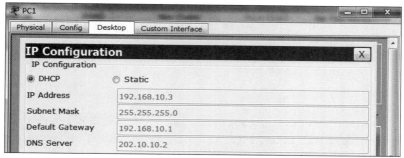

图 13-4　PC1 自动获取地址结果

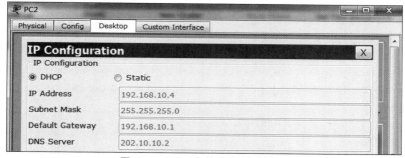

图 13-5　PC2 自动获取地址结果

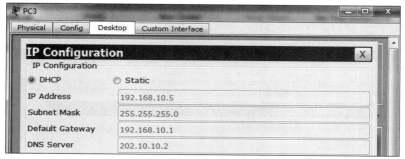

图 13-6　PC3 自动获取地址结果

通过上面的实验,可以看到 PC1、PC2 和 PC3 通过网络向 DHCP 服务器发送请求,获取到了 IP 地址。

3. 路由器的配置

本实验,将 WWW、DNS、FTP 服务分别部署在不同的服务器中,并且使它们属于不同的网段,所以在进行各服务器的搭建和测试时,首先要保证网络的连通。在此,先进行路由

器的配置,实现网络的互通,命令内容如下:

```
Router>enable
Router#configure terminal
Enter configuration commands, one per line. End with CNTL/Z.
Router(config)#interface g0/2
Router(config-if)#no shutdown
Router(config-if)#
%LINK-5-CHANGED: Interface GigabitEthernet0/2, changed state to up
Router(config-if)#ip address 192.168.10.1 255.255.255.0
Router(config-if)#exit
Router(config)#interface g0/1
Router(config-if)#no shutdown
Router(config-if)#
%LINK-5-CHANGED: Interface GigabitEthernet0/1, changed state to up
%LINEPROTO-5-UPDOWN: Line protocol on Interface GigabitEthernet0/1, changed state to up
Router(config-if)#ip address 202.10.11.1 255.255.255.0
Router(config-if)#exit
Router(config)#interface g0/0
Router(config-if)#no shutdown
Router(config-if)#
%LINK-5-CHANGED: Interface GigabitEthernet0/0, changed state to up
%LINEPROTO-5-UPDOWN: Line protocol on Interface GigabitEthernet0/0, changed state to up
Router(config-if)#ip address 202.10.10.1 255.255.255.0
Router(config-if)#exit
```

4. FTP 服务器的搭建

1) FTP 服务器的地址规划

此处,模拟 FTP 服务来自公网,所以,实验中给该服务器配置一个公网地址,如图 13-7 所示。

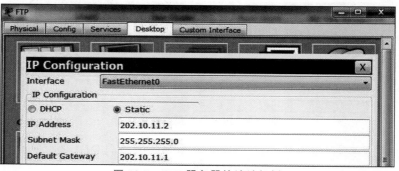

图 13-7　FTP 服务器的地址规划

2) FTP 服务器的配置与实现

打开服务器配置界面,创建用户名和密码,分配权限。FTP 服务器的配置如图 13-8 所示。

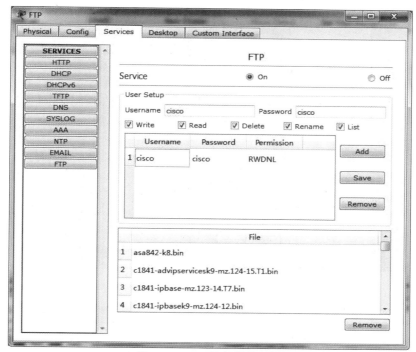

图 13-8　FTP 服务器的配置

3）FTP 服务器的测试

（1）FTP 服务器的登录测试。打开网络中任意一台计算机的 MS-DOS 环境，此处以 PC3 为例，进行 FTP 服务器访问测试，在其 MS-DOS 命令行中输入 ftp 202.10.11.2，按 Enter 键就会访问 FTP 服务器，且需要输入用户名和密码，登录成功后，就会到达 FTP 服务器的根目录。FTP 服务器的登录测试如图 13-9 所示。

图 13-9　FTP 服务器的登录测试

（2）FTP 文件服务器中文件的查看。登录 FTP 服务器后，输入命令 dir，按 Enter 键查看 FTP 服务器中的文件，如图 13-10 所示。

5. WWW 服务器的构建

1）WWW 服务器的地址规划

WWW 服务是目前应用最广的一种基本互联网应用，人们每天上网都要用到这种服务，在本次实验中，进行 WWW 服务的构建，它的地址规划如图 13-11 所示。

2）WWW 服务器的配置

开启超文本传输协议，设置主页信息，构建网站，对 WWW 服务器的配置如图 13-12 所示。

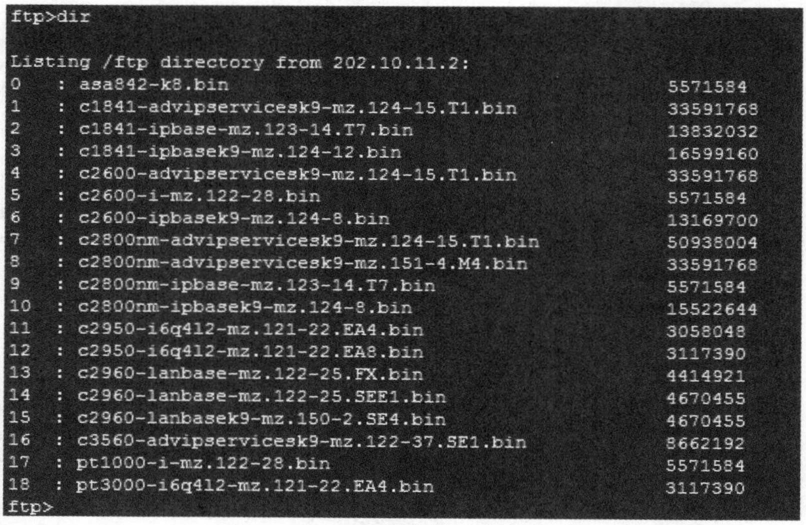

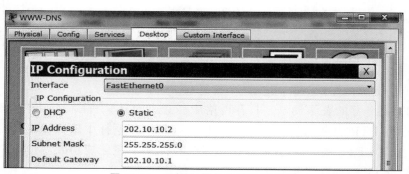

图 13-10　FTP 服务器文件的查看

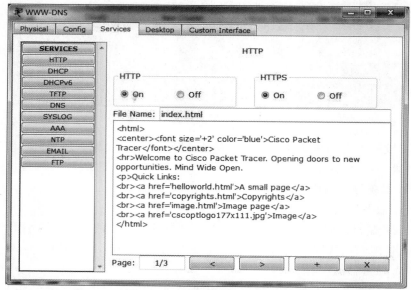

图 13-11　WWW 服务器的地址规划

图 13-12　WWW 服务器的配置

3）WWW 服务器的测试

本实验，利用 PC3 来进行 WWW 服务器的测试。打开 PC3 的浏览器，在地址栏中输入 http://202.10.10.2，按 Enter 键后，就能打开主页。WWW 服务器的测试如图 13-13 所示。

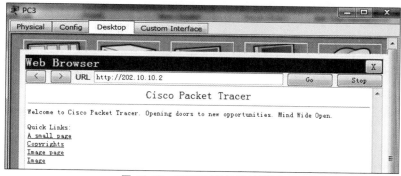

图 13-13　WWW 服务器的测试

6. DNS 服务器的构建

1）DNS 服务器的地址规划

本次实验，将 DNS 和 WWW 服务部署在同一个服务器上，所以直接在 WWW 服务器上进行 DNS 服务器的构建，地址保持不变。

2）DNS 服务器的配置

开启 DNS 服务，添加域名及解析地址，DNS 服务器的构建如图 13-14 所示。

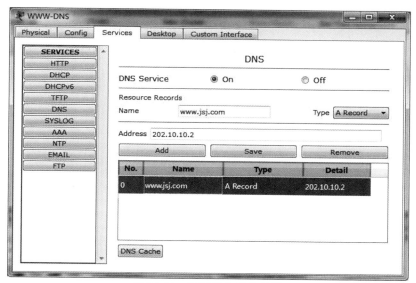

图 13-14　DNS 服务器的构建

3）DNS 服务器的测试

打开 PC3 的浏览器，在地址栏中输入 http://www.jsj.com，按 Enter 键就可以打开主页，DNS 服务器的测试如图 13-15 所示。

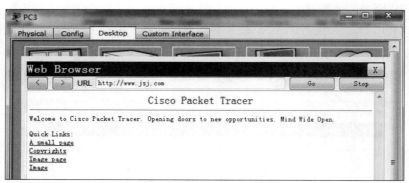

图 13-15　DNS 服务器的测试

五、实验总结

本实验让学生熟悉了 DHCP、FTP、WWW 和 DNS 服务的功能，并掌握了各种服务的构建，提高了学生对网络的理解，加深了对网络服务的认识和应用能力。

六、实验思考题

（1）在同一 WWW 服务器上能否建立多个 Web 站点？若能建立，在配置时有哪些注意事项？

（2）简要说明 FTP、DNS、DHCP、WWW 服务的基本功能。

实验十四　小型无线网络的设计与配置

无线网是一种重要的网络接入技术,目前已经得到广泛的应用,在众多的无线网络协议中,最重要的标准是 IEEE 802.11。无线网络是指在没有布线的情况下,就能实现各种通信设备互连的网络。无线网络技术涵盖的范围很广,既包括允许用户建立远距离无线连接的全球语音和数据网络,也包括为近距离无线连接进行优化的红外线及射频技术。根据网络覆盖范围的不同,可以将无线网络划分为无线广域网(wireless wide area network,WWAN)、无线局域网(wireless local area network,WLAN)、无线城域网(wireless metropolitan area network,WMAN)和无线个人区域网(wireless personal area network,WPAN)。

小型无线网络的设计与配置

与有线网络相比,无线网络的主要特点是完全消除了有线网络的局限性,实现了信息的无线传输,使人们更自由地使用网络。同时,网络运营商操作起来也非常方便,体现在线路建设成本降低,运行时间缩短,成本回报和利润生产相对较快等。同时包括改进了管理员的无线信息传输管理方式,也为网络中没有空间限制的用户提供了更大的灵活性。

一、实验目的

(1) 熟悉无线网络的概念,了解无线 AP、胖 AP 以及瘦 AP 的区别。
(2) 掌握无线局域网的基本组成以及设备互连方式。
(3) 掌握无线路由器的配置方法。
(4) 掌握小型无线网络的设计与实现方法。

二、实验环境

Cisco Packet Tracer 模拟器(服务器、计算机、笔记本电脑、网线)。

三、实验内容及步骤

(1) 无线 AP、胖 AP 以及瘦 AP 介绍。
(2) 在模拟器中构建无线局域网。
(3) 无线网络的测试。
(4) 家庭无线网络的配置。

四、实验过程

1. 无线 AP、胖 AP 以及瘦 AP 介绍

无线 AP(access point)，即无线访问结点，也叫无线接入点，可以理解成最末端的无线交换机，负责接入无线网卡，如笔记本电脑、手机、平板。它分为胖 AP 和瘦 AP 两种。

1) 胖 AP

胖 AP 一般还同时具备 WAN、LAN 端口，支持 DHCP 服务器、DNS 和 MAC 地址克隆、VPN 接入、防火墙等安全功能。它通常自带完整的操作系统，是可以独立工作的网络设备，可以实现拨号、路由等功能，一个典型的例子就是生活中常见的无线路由器。胖 AP 一般用于小型无线网络建设，无须 AC 协作即可独立工作。一般用于家庭、小型企业或小型办公场景，只需要少量的完整覆盖。但是它无法实现无线漫游，用户从一个胖 AP 的覆盖区域移动到另一个胖 AP 的覆盖区域，需要重新连接胖 AP 信号，重新认证，重新获取 IP 地址，从而导致网络断开连接现象频发。当许多用户连接到同一个胖 AP 时，它无法自动执行负载均衡，无法将用户分配给负载较轻的其他胖 AP。因此，胖 AP 可能会因重负载而出现网络故障。胖 AP 无法集中管理，它们需要逐个单独配置，配置工作烦琐。

2) 瘦 AP

瘦 AP 删除了路由、DNS 和 DHCP 服务器以及许多其他加载功能，只保留无线接入部分。人们经常提到的 AP 指的就是这样的瘦 AP，其相当于无线交换机或集线器，并且仅提供一种有线/无线信号转换和无线信号接收/发送功能。作为无线局域网的一个组成部分，瘦 AP 无法独立工作，因为它没有管理功能，它必须与无线控制器(AC)合作才能成为一个完整的系统，它仅负责广播 SSID 和连接终端。它通常用于中型和大型无线网络的构建，例如一定数量的 AP 与 AC 产品合作，可以形成更大的无线网络覆盖，使用场景通常是购物中心、超市、景点、酒店、公司等。当用户从一个瘦 AP 的覆盖区域移动到另一个瘦 AP 的覆盖区域时，信号自动切换，无须重新认证，无须重新获取 IP 地址，网络始终在线连接，使用方便。当多个用户连接到同一个瘦 AP 时，AC 会根据负载均衡算法自动将用户分配给负载较轻的其他 AP，从而降低 AP 的故障率，提高 AP 的性能。

虽然瘦 AP 有很多优点，但是基于成本，很多家庭、餐馆、咖啡厅、4S 店、旅馆、美容院、健身房等都选择使用多种胖 AP 来搭建无线网络。

2. 在模拟器中构建无线局域网

设备选择：1841 系列路由器 1 台、WRT300N 系列路由器 1 台、计算机 2 台、笔记本电脑 1 台、服务器 1 台。

通信线缆：交叉线 2 根、直通线 1 根。

端口连接：服务器 Server0 的 F0 接口连接 1841 Router0 的 f0/1 接口，1841 Router0 的 f0/0 接口连接无线路由器的 Internet 接口，无线路由器的 Ethernet 1 接口连接 PC1 的 F0 接口，PC0 和笔记本电脑 Laptop0 通过无线连接，如图 14-1 所示。

1) 地址规划

IP 地址规划如表 14-1 所示。

实验十四 小型无线网络的设计与配置

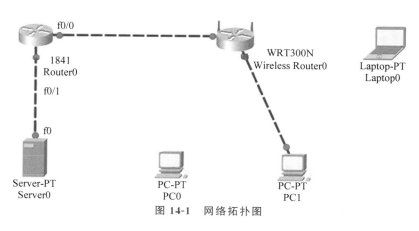

图 14-1 网络拓扑图

表 14-1 IP 地址规划表

序号	设 备	接 口	IP 地址	子网掩码
1	Server0	FastEthernet0	10.0.0.100	255.255.255.0
2	Router0	FastEthernet 0/1	10.0.0.1	255.255.255.0
3	Router0	FastEthernet 0/0	11.0.0.1	255.255.255.0
4	Wireless Router0	Internet	11.0.0.2	255.255.255.0
5	Wireless Router0	网关	11.0.0.1	255.255.255.0

2) 设备基本配置

(1) 服务器的配置。参照表 14-1 的地址规划信息,打开服务器配置界面,选择 Config 标签,选择 FastEthernet0 选项,配置服务器的 IP 地址为 10.0.0.100,如图 14-2 所示。选择 Global→Settings 选项,配置服务器的网关地址为 10.0.0.1,如图 14-3 所示。

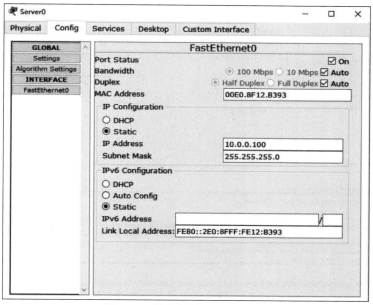

图 14-2 服务器的 IP 地址配置

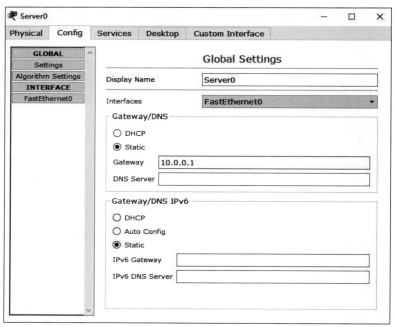

图 14-3　服务器的网关地址

（2）1841 路由器配置。打开路由器的配置界面，选择 Config 标签，选择 FastEthernet0/1 选项，选择 Port Status 后面的 On 单选按钮，表示开启该端口，配置该端口的 IP 地址为 10.0.0.1，如图 14-4 所示。

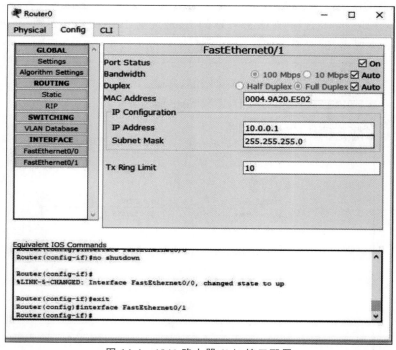

图 14-4　1841 路由器 f0/1 接口配置

选择 FastEthernet0/0 选项，选择 Port Status 后面的 On 单选按钮，表示开启该端口，配置该端口的 IP 地址为 11.0.0.1，如图 14-5 所示。

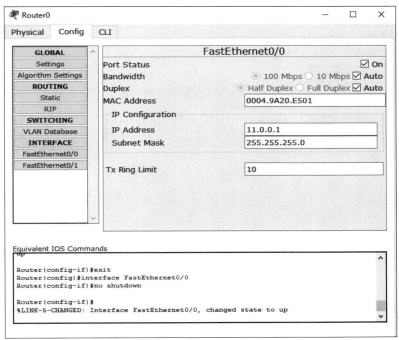

图 14-5　1841 路由器 f0/0 接口配置

（3）无线路由器 Wireless Router0 的配置。打开路由器的配置界面，选择 Config 标签，选择 INTERFACE→Internet 选项，配置互联网接口。将 IP 配置为静态，设置默认网关为 11.0.0.1，IP 地址为 11.0.0.2，如图 14-6 所示。

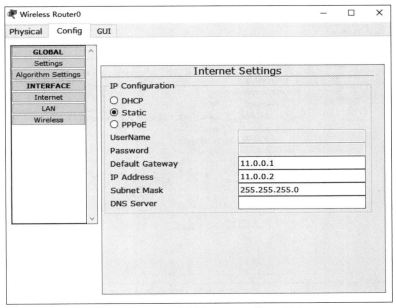

图 14-6　无线路由器 Internet 接口配置

接下来配置无线路由器的本地局域网(LAN),本地局域网有有线和无线两种连接方式,本实验对两种方式都进行实践操作。LAN 使用默认 IP 地址:192.168.0.1,如图 14-7 所示。

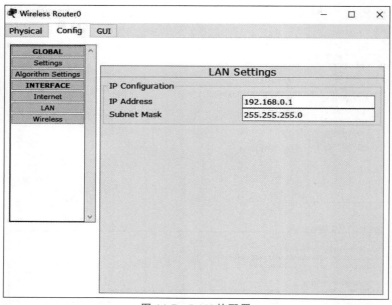

图 14-7　LAN 的配置

随后,配置无线路由器的无线接入端口,修改 SSID 为 jsjx。SSID 为无线路由器提供无线网络账号,认证方式选择 WPA2-PSK 单选按钮,密码为 cisco123。所有无线终端,都需要通过指定的认证才能接入,且需要输入密码 12345678,如图 14-8 所示。

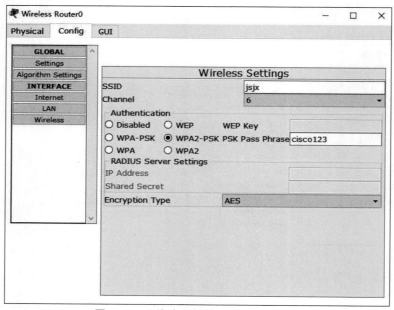

图 14-8　无线路由器的账号和认证配置

（4）计算机无线网卡的装载。为 PC0 和 Laptop0 装载无线网卡，在 Cisco Packet Tracer 模拟器中，需要先卸载原有线网卡，然后才能装载无线网卡，具体过程：关闭电源→拖出原有线网卡→拖入无线网卡→打开电源，如图 14-9 所示。

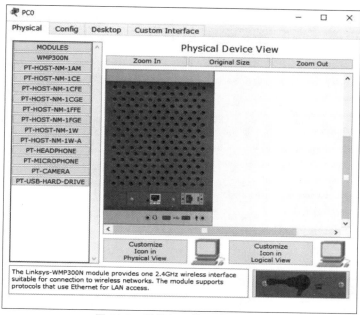

图 14-9　计算机无线网卡的装载

对 PC0 进行上网认证，输入正确的 SSID 和 WPA2-PSK，如图 14-10 所示。

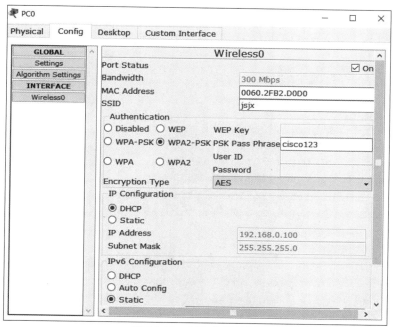

图 14-10　PC0 上网认证

设置 PC0 的 IP 为自动获取(DHCP)模式,可以看到 PC0 获得了合法的 IP 地址,如图 14-11 所示。

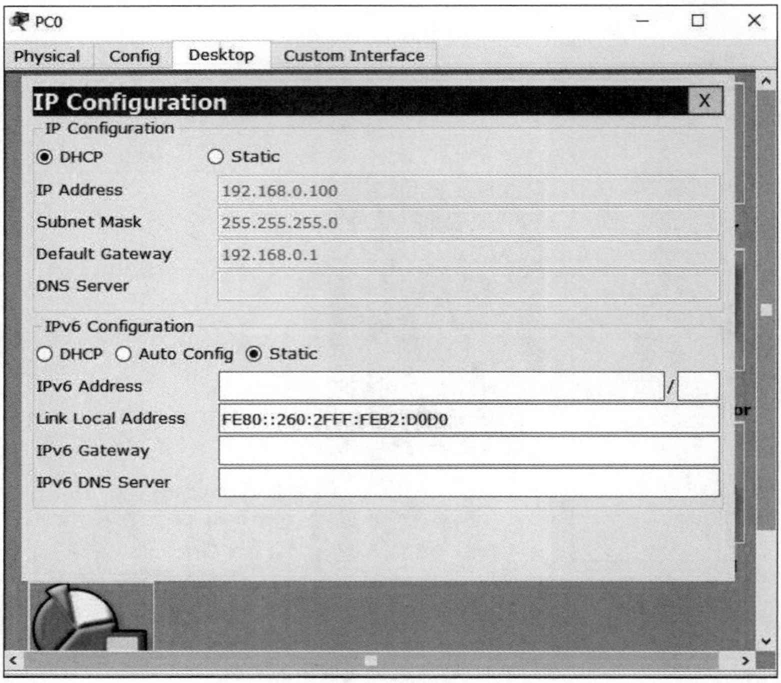

图 14-11　PC0 自动获取 IP 地址

按照同样的方法,PC1 和 Laptop0 也能获取到合法地址。

3. 网络测试

用 PC0 去 ping 无线路由器的 IP 地址,命令内容如下:

```
Packet Tracer PC Command Line 1.0
PC>ping 11.0.0.2

Pinging 11.0.0.2 with 32 bytes of data:

Reply from 11.0.0.2: bytes=32 time=0ms TTL=255
Reply from 11.0.0.2: bytes=32 time=0ms TTL=255
Reply from 11.0.0.2: bytes=32 time=0ms TTL=255
Reply from 11.0.0.2: bytes=32 time=0ms TTL=255

Ping statistics for 11.0.0.2:
Packets: Sent = 4, Received = 4, Lost = 0 (0% loss),
Approximate round trip times in milli-seconds:
Minimum = 0ms, Maximum = 0ms, Average = 0ms
```

从得到的信息可以看出 PC0 到网关是连通的,接下来测试 PC0 和外网 Web 服务器的连通性,直接用 PC0 去 ping 10.0.0.100,命令内容如下:

```
PC>ping 10.0.0.100

Pinging 10.0.0.100 with 32 bytes of data:

Reply from 10.0.0.100: bytes=32 time=0ms TTL=254
Reply from 10.0.0.100: bytes=32 time=0ms TTL=254
Reply from 10.0.0.100: bytes=32 time=0ms TTL=254
Reply from 10.0.0.100: bytes=32 time=0ms TTL=254

Ping statistics for 10.0.0.100:
Packets: Sent = 4, Received = 4, Lost = 0 (0% loss),
Approximate round trip times in milli-seconds:
Minimum = 0ms, Maximum = 0ms, Average = 0ms
```

从得到的信息可以看出，PC0 与外网是能够通信的，可以通过 PC0 的浏览器访问 Web 服务器，如图 14-12 和图 14-13 所示。

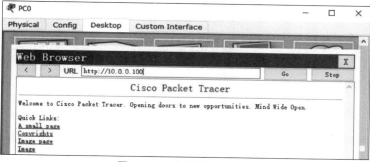

图 14-12　访问外部网站 1

图 14-13　访问外部网站 2

4. 家庭无线网络的配置

（1）结合家庭的环境和布局，选择合适的无线路由器，主要考虑性能、功率、覆盖范围、芯片以及品牌等。无线路由器上有一个黄颜色的接口，也叫 WAN 口，就是人们常说的外网接口，同时一般还配有4个LAN口，即局域网接口，可用有线的方式来连接4个终端，一般为蓝色。不同的路由器，对应的接口的传输速率是不一样的，一般有100Mbps或1000Mbps，可结合需求选择。每台路由器的标签上都有管理地址，可通过该地址进行路由器的管理和配置，如图14-14所示。

图14-14　无线路由器的接口和管理地址

（2）将接入网的总线接到无线路由器的WAN口，将计算机通过网线连接到任意一个LAN口，打开计算机的浏览器，在其地址栏中输入路由器的管理地址，就会进入路由器的管理界面，本次实验使用的路由器管理地址为http://tplogin.cn，如图14-15所示。

图14-15　路由器的管理界面

（3）进行路由器的配置，选择"设置向导"选项，如图14-16所示。

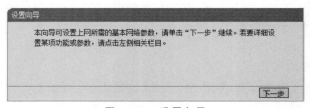

图14-16　设置向导

(4) 单击"下一步"按钮,进行上网方式的设置,选择"让路由器自动选择上网方式(推荐)"单选按钮,如图 14-17 所示。

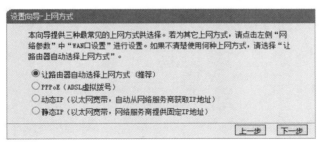

图 14-17　选择路由器的上网方式

(5) 单击"下一步"按钮,路由器会自动检查当前的网络环境,大概需要 5s 的时间,如图 14-18 所示。

图 14-18　自动检查网络环境

(6) 当前的网络环境是静态 IP 方式,所以需要输入网络运营商提供的固定 IP 地址,如图 14-19 所示。如果是拨号上网的方式,此处就会弹出输入网络运营商提供的账号信息。

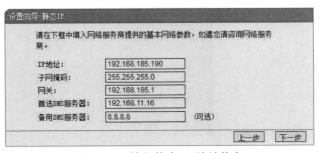

图 14-19　输入静态 IP 地址信息

(7) 单击"下一步"按钮,设置无线网络信息,自定义 SSID(账号)以及安全认证,如图 14-20 所示。

(8) 设置完成后,单击"下一步"按钮,如图 14-21 所示。至此,单击"完成"按钮,无线配置完成。

(9) 还可以通过主界面左侧的"网络参数"选项进行相关信息的修改和完善,如图 14-22 所示。

(10) 接入网络类型、地址、网关、DNS 的配置和修改,如图 14-23 所示。

(11) 进行 LAN 接口设置,如图 14-24 所示。

(12) 进行无线网络安全的设置,如图 14-25 所示。

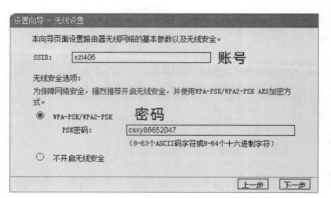

图 14-20　无线网络设置

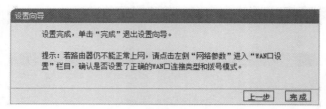

图 14-21　无线网络设置完成

图 14-22　网络参数选项

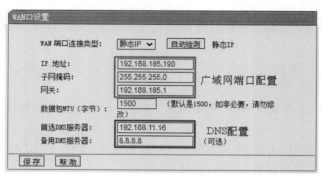

图 14-23　WAN 端口的设置

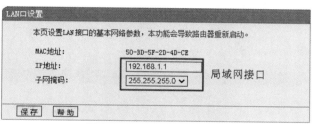

图 14-24　LAN 接口的设置

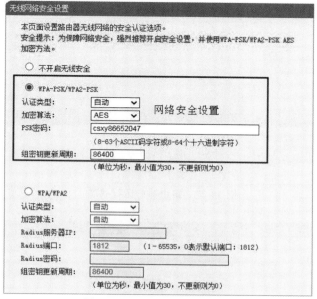

图 14-25　无线网络安全设置

（13）进行 DHCP 设置，通过该项的设置，可以确定地址池的范围大小，如图 14-26 所示。

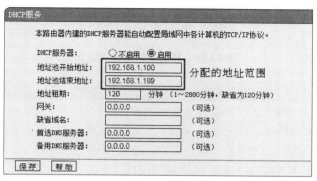

图 14-26　DHCP 设置

（14）设置完成后，再次查看路由器的运行状态，如图 14-27 和图 14-28 所示。

（15）还可以查看连接到该路由器上的所有终端信息。可以看到该路由器一共连接了 3 台设备，以及每台设备接收和发送数据包的大小，如图 14-29 所示。

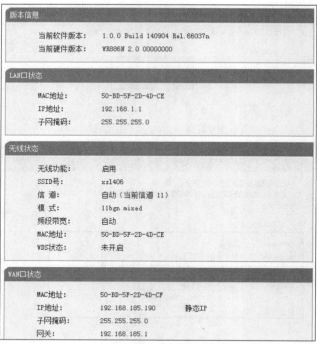

图 14-27 路由器的运行状态 1

图 14-28 路由器的运行状态 2

图 14-29 连接终端信息

五、实验总结

本实验让学生更好地了解无线网,掌握无线网组建的方法和技术,掌握无线路由器的作用及应用,熟悉无线适配器的用途和应用方法,熟悉 IP 地址的获取方式,掌握如何选择网络设备和通信线缆,掌握构建小型无线网络的方法,掌握网络连通的各种测试方法。

六、实验思考题

（1）本实验中 PC0 和 PC1 获取地址的途径有什么区别？

（2）本实验中所选择的无线路由器最多可以通过有线连接的方式为多少台终端提供上网服务？

实验十五　网络打印服务的配置

随着网络技术的快速发展和应用,网络已经渗透到了人们生活、学习和工作的各个领域,使人们的生活更加丰富多彩,生活的质量也在不断提高。近年来,人们对网络打印服务的需求不断扩大,然而,对于网络打印服务所需的配置,仍有很多人不了解,在设备出现故障时无法解决,直接影响了学习、工作的效率。本实验从实际应用出发,从网络共享打印服务以及网络打印机的配置两方面讲解,过程详尽,便于读者理解和掌握。

一、实验目的

(1) 熟悉网络环境和打印服务。
(2) 掌握本地打印机的安装方式。
(3) 掌握共享打印服务的配置方法。
(4) 掌握网络打印机的配置方法。

二、实验环境

局域网、计算机 2 台、家用打印机(HP LaserJet 1020)1 台、网络打印机(LaserJet Pro MFP M226dn)1 台。

三、实验内容及步骤

(1) 本地打印机的安装。
(2) 网络共享打印机的配置。
(3) 网络打印机的配置。

四、实验过程

1. 本地打印机的安装

1) 连接打印机与计算机

通过数据线将打印机与计算机连接起来,接通打印机的电源,计算机会显示发现新硬件,随即进行自动安装。但这一过程往往都会因为系统缺乏合适的驱动程序,导致该硬件安装失败,此时系统提示:"硬件无法正常使用"。

2) 进行打印机驱动程序的安装

一般获取打印机的驱动程序有 3 种方式:①购买打印机时,会随设备附有一张光盘，内有该打印机的驱动程序；②通过 360、软件大师等第三方软件进行下载；③通过浏览器搜索或进入官方网站下载。安装驱动程序，双击运行 SETUP.exe 文件，如图 15-1 所示。

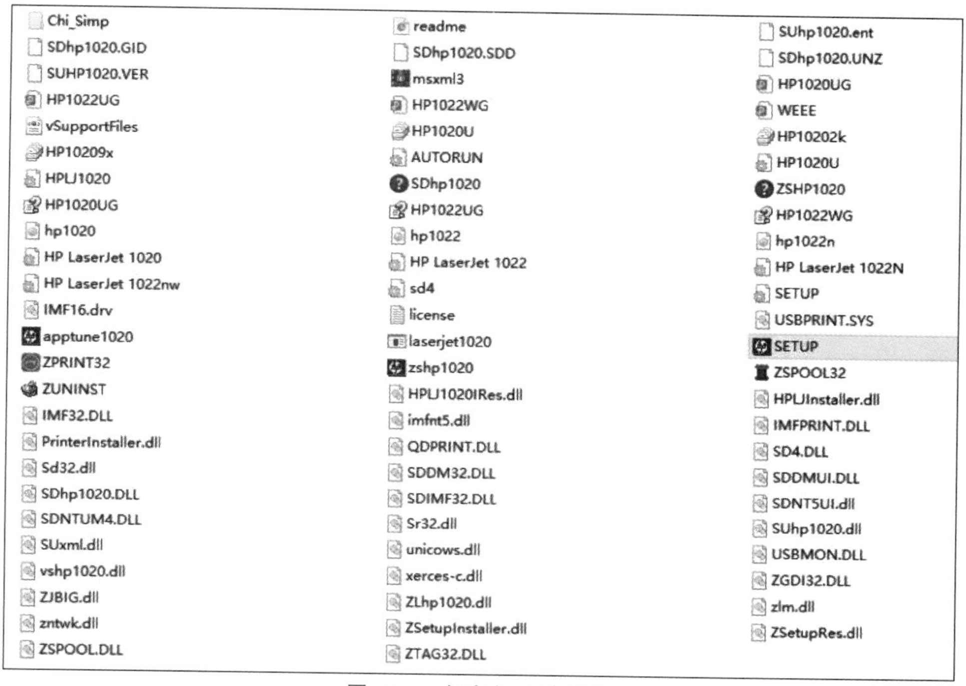

图 15-1　打印机驱动安装

3）打印机测试

根据安装向导的提示，选择默认安装方式，直到安装结束，会提示安装成功，选中"打印测试页"复选框，单击"完成"按钮，系统会自动启动打印服务，进行测试页的打印，如图 15-2 所示。

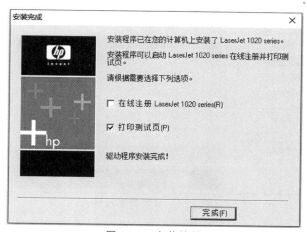

图 15-2　安装结果

通过打印的测试页，可以看到计算机的名称、打印机的名称、打印机的型号等，这些信息有助于进行网络共享打印服务的配置，如图15-3所示。

图15-3　打印机信息

2. 网络共享打印机的配置

1) 设置默认打印机

单击"开始"按钮，选择"设备和打印机"选项，双击安装好的打印机"HP LaserJet 1020 (副本1)"，打开打印机的"打印队列"对话框，打开"打印机"菜单，选择"设置为默认打印机"菜单项(完成该操作后，在本机的各种软件中，只要单击"打印"按钮，该打印机就会自动响应，进行资料的打印)。再选择"共享"菜单项，如图15-4所示。

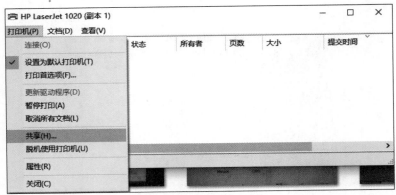

图15-4　启动共享设置

2) 共享打印机配置

在弹出的"打印机属性"对话框中选择"共享"标签，打开共享设置选项卡，选中"共享这台打印机"以及"在客户端计算机上呈现打印作业"复选框,输入共享名(该名称可以自行拟定,系统默认就是打印机的名称),如图 15-5 所示。

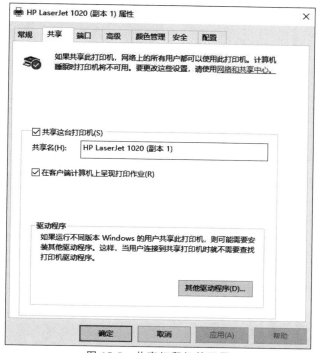

图 15-5　共享打印机的配置

单击"应用"或"确定"按钮,共享打印机就设置好了。

3) 客户机添加打印机

客户机添加打印机时一定要注意:该计算机必须和刚才设置共享打印机的那台计算机在同一个局域网中。

(1) 单击"开始"→"运行"按钮,或按快捷键 Windows＋R,打开"运行"对话框,输入\\192.168.1.104,如图 15-6 所示。

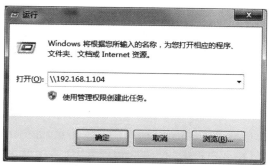

图 15-6　"运行"对话框

(2) 单击"确定"按钮后,弹出"Windows 安全"对话框,输入共享了打印机的那台计算机的登录账号和密码,进行权限认证,如图 15-7 所示。

图 15-7　权限认证

（3）选中"记住我的凭据"复选框，以后再次访问时，就不需要认证了，单击"确定"按钮，就会打开 192.168.1.104 这台计算机的共享信息窗口，如图 15-8 所示。

图 15-8　共享信息窗口

（4）在共享的信息窗口中，可以看到共享的打印机，双击该打印机的图标就会进行安装，根据提示进行操作，大概 5s 后就会安装成功，在弹出的对话框中单击"打印测试页"按钮，结果如图 15-9 所示。

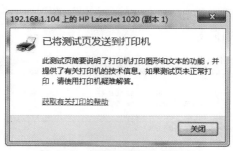

图 15-9　打印测试页执行信息

（5）从打印的测试信息中，可以看到打印机的基本信息，如图 15-10 所示。

（6）打印机添加完成，在客户机的"打印机和传真"列表中就可以看到添加的网络打印机设备，如图 15-11 所示。同样，将该打印机设置为"默认打印机"。

下面再介绍一种添加打印机的方式。

（1）打开"设备和打印机"界面，单击"添加打印机"按钮，弹出"添加打印机"对话框，如图 15-12 所示。

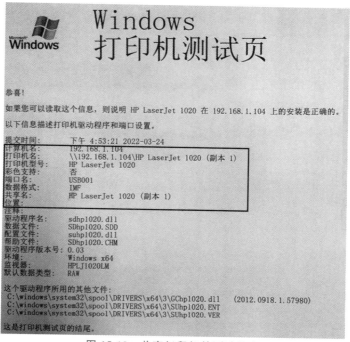

图 15-10　共享打印机的测试信息

图 15-11　添加的网络打印机

（2）选择"添加网络、无线或 Bluetooth 打印机"选项，开始搜索网络中的打印机，如图 15-13 所示。

（3）如果需要添加的打印机不在队列中，就单击"我需要的打印机不在列表中"按钮，然后，单击"下一步"按钮，在弹出的对话框中，选择"按名称选择共享打印机"单选按钮，并输入正确的网络共享打印机的名称后，单击"下一步"按钮，如图 15-14 所示。

（4）成功添加网络共享打印机，如图 15-15 所示。

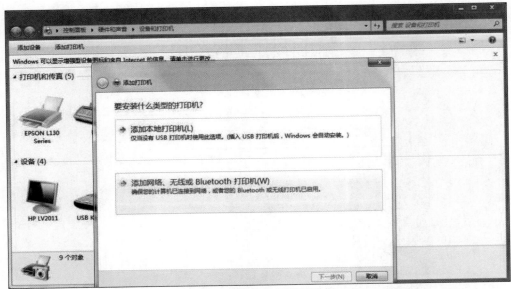

图 15-12　添加打印机对话框

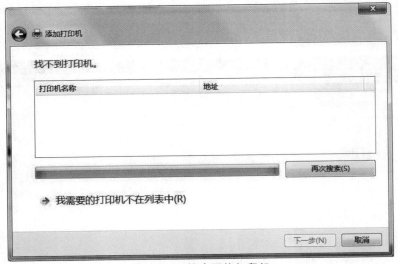

图 15-13　搜索网络打印机

（5）单击"下一步"按钮，打印测试页，验证结果，如图 15-16 所示。

最后，还要说明一下，如果出现图 15-17 所示的现象，说明有文档正在打印。当有多个文档都执行了打印命令后，就会在该窗口进行排队，按顺序打印。不过，有时显示"正在打印"状态，但是打印机没有任何反应，说明打印服务故障，一般需要重新添加打印机或者重新安装驱动程序后再添加打印机。

3. 网络打印机的配置

网络打印机要接入网络，一定要有网络接口。目前有两种接入的方式，一种是打印机自带打印服务器，打印服务器上有网络接口，只需插入网线，分配 IP 地址就可以了；另一种是打印机使用外置的打印服务器，打印机通过并口或 USB 口与打印服务器连接，打印服务器

实验十五　网络打印服务的配置

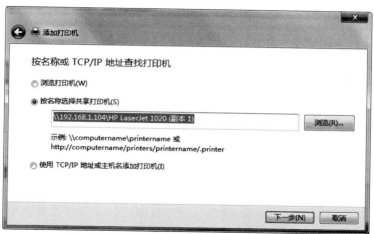

图 15-14　输入网络共享打印机的名称

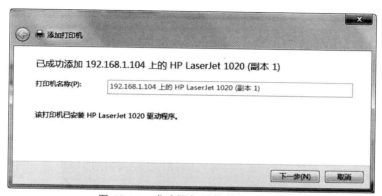

图 15-15　成功添加网络共享打印机

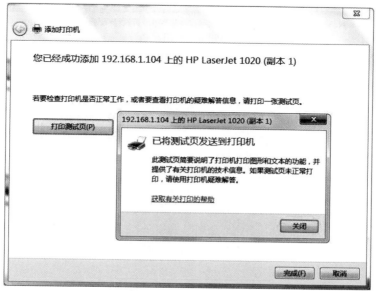

图 15-16　打印测试页

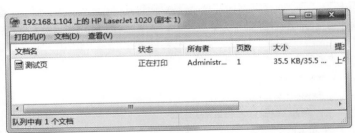

图 15-17　打印序列

再与网络连接。网络打印机一般具有管理和监视软件，通过管理软件可以远程查看和干预打印任务，对打印机的配置参数进行设定，绝大部分的网络打印管理软件都是基于 Web 方式的。而通过管理软件，用户可以查看打印任务、打印机的工作状态等信息。一般来说，管理软件是给网络管理员或者高级用户使用的，而普通用户则可以使用打印机的监视功能。网络打印机用于网络系统，要为多数人提供打印服务，因此要求这种打印机具有打印速度快、能自动切换仿真模式和网络协议，便于网络管理员进行管理等优势。

下面以网络打印机 LaserJet Pro MFP M226dn 为例，讲解网络打印机的配置。

（1）将打印机接入局域网，打开设置菜单，选择网络设置，如图 15-18 所示。

图 15-18　网络设置

（2）进行网络配置，选择 IPv4 配置方法，如图 15-19 所示，下翻选择"手动输入"选项。

图 15-19　选择 IPv4 配置方法

（3）设置 IP 地址，手动输入 169.254.036.121（注意：该打印机默认 IP 地址每段都是 3 位的，所以 36 前面要补 0），如图 15-20 所示。

图 15-20　输入 IP 地址

（4）打印机设置完成，如图 15-21 所示。

图 15-21　打印机设置完成

（5）打开局域网中的计算机，打开"添加打印机"对话框，选择"使用 TCP/IP 地址或主机名添加打印机"单选按钮，再单击"下一步"按钮，如图 15-22 所示。

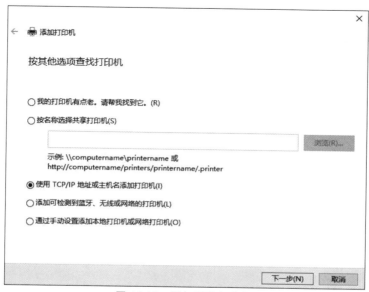

图 15-22　添加网络打印机

(6)输入网络打印机的 IP 地址,端口名称默认与 IP 地址相同,如图 15-23 所示。

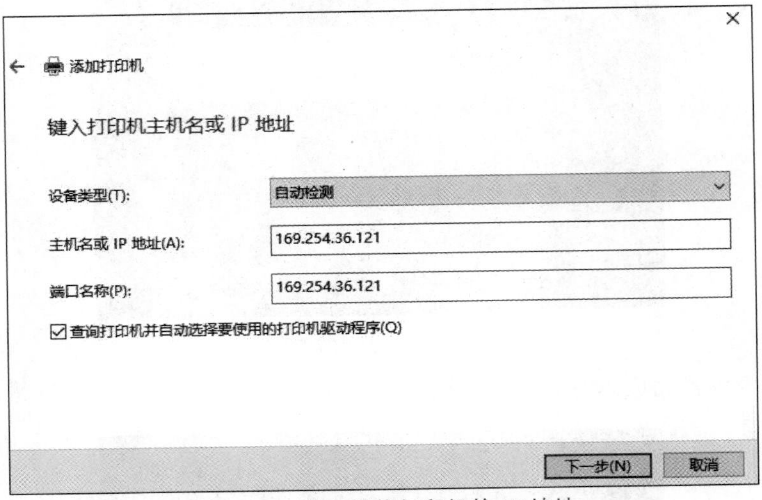

图 15-23　输入网络打印机的 IP 地址

(7)单击"下一步"按钮,系统会进行网络打印机的安装,其过程以及测试方法与前面的内容相同,此处不再赘述。

五、实验总结

本实验让学生掌握了打印机的安装方法、网络共享打印机的配置方法以及网络打印机的配置方法,实现了网络打印服务,方便了日常办公。实验通过多种方法实现打印服务,既让学生掌握了应用技能,也指导学生从不同的角度思考问题,对其解决实际问题的能力有很大促进作用,特别是对以后的就业和工作有很大的帮助。

六、实验思考题

(1)打印机的驱动程序有什么作用?
(2)在局域网中实现网络打印服务的条件有哪些?

技能提升篇

实验十六 生成树协议的原理与验证

为了保证网络的稳定性,减少网络故障时间,通常情况下可以使用冗余拓扑,而冗余网络的引入会产生交换环路的情况。伴随着交换环路的3个问题是广播风暴、同帧的复制以及交换机的CAM表结构振荡。STP则可以在冗余结构网络环境下发生故障时自行转换数据的转发路径。STP(spanning tree protocol,生成树协议)能够处理好交换环路产生的很多问题,STP基础性思路是通过阻断某些交换机自身端口,构造一棵树,即无环路的可以转发的树。STP通过BPDU(bridge protocol data unit)报文在交换之间进行协商,确定需被阻塞的端口。STP有很多版本,分别支持不同的标准,早期的IEEE 802.1d重新收敛时间较长,需要30~50s,为了减少收敛时间,Cisco公司引入了补充技术,如UplinkFast、BackboneFast和PortFast等;RSTP(rapid spanning tree protocol,快速生成树协议)(IEEE 802.1w)则在协议底层更新形成新协议,通过这种方式减少收敛的时间;而PVST+和MSTP(IEEE 802.1s)都是STP的改进版本。

一、实验目的

本实验模拟RSTP实验创建的前后变化,并测试线路冗余,了解设置主根之后产生的影响,掌握生成树协议的原理和技术。

二、实验环境

Cisco Packet Tracer 6.0模拟器(交换机、计算机、直通线、Console配置线)、2950系列交换机1台、2960系列交换机3台、3560交换机2台、计算机2台、服务器2台。

三、实验内容及步骤

(1) 拓扑搭建。
(2) 进行VLAN和接口等二层基础信息的配置。
(3) 配置生成树。
(4) 网络测试。
(5) 冗余连接的应用。
(6) 查看状态信息。

四、实验过程

1. 实验拓扑

端口连接：1 台 2950T-24 交换机与 3 台 2960-24TT 交换机分别通过 f0/23 和 f0/24 接口互连，并通过 G0/1 和 G0/2 接口连接到 3560 交换机的 G0/1、G0/2、f0/1、f0/2 接口；PC0 和 PC1 连接到 Switch3 和 Switch5 的 f0/1 接口；Server0 和 Server1 连接到 Switch4 和 Switch6 的 f0/1 接口，拓扑如图 16-1 所示。

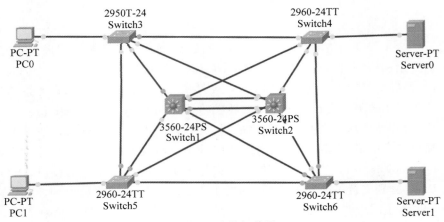

图 16-1　网络拓扑图

结合实验的内容，IP 地址规划如表 16-1 所示。

表 16-1　IP 地址规划表

序　号	设　备	接　口	IP 地址	子网掩码
1	PC0	FastEthernet0	192.168.10.1	255.255.255.0
2	PC1	FastEthernet0	192.168.10.2	255.255.255.0
3	Server0	FastEthernet0	192.168.20.1	255.255.255.0
4	Server1	FastEthernet0	192.168.20.2	255.255.255.0

2. 实验配置

Switch1 配置封装协议 dot1q，接口配置为 trunk 模式，如图 16-2 所示。

```
Switch1(config)#vlan 10
Switch1(config-vlan)#vlan 20
Switch1(config-vlan)#int r f0/1-4
Switch1(config-if-range)#sw tr en dot
Switch1(config-if-range)#sw mo tr
Switch1(config)#ip routing
Switch1(config)#int vlan 10
Switch1(config-if)#ip add192.168.10.254 255.255.255.0
```

图 16-2　Switch1 配置

同样，将 Switch2 也配置封装协议 dot1q，接口配置为 trunk 模式，如图 16-3 所示。

```
Switch2(config)#vlan 10
Switch2(config-vlan)#vlan 20
Switch2(config-vlan)#int r f0/1-4
Switch2(config-if-range)#sw tr en dot
Switch2(config-if-range)#sw mo tr
Switch2(config)#ip routing
Switch2(config)#int vlan 20
Switch2(config-if)#ip add 192.168.20.254 255.255.255.0
```

图 16-3　Switch2 配置

Switch3 配置接口为 access 模式,并将接口 f0/1 划分到 VLAN 10 中,配置如图 16-4 所示。

Switch4 的接口 f0/23、f0/24、G0/1、G0/2 配置为 trunk 模式,并将 f0/1 接口划分到 VLAN 20 中,如图 16-5 所示。

```
Switch3(config)#vlan 10
Switch3(config-vlan)#vlan 20
Switch3(config-vlan)#int r f0/23-24
Switch3(config-if-range)#sw mo tr
Switch3(config-if-range)#int r g0/1-2
Switch3(config-if-range)#sw mo tr
Switch3(config-if-range)#int f0/1
Switch3(config-if)#sw mo acc
Switch3(config-if)#sw acc vlan 10
```

图 16-4　Switch3 配置

```
Switch4(config)#vlan 10
Switch4(config-vlan)#vlan 20
Switch4(config-vlan)#int r f0/23-24
Switch4(config-if-range)#sw mo tr
Switch4(config-if-range)#int r g0/1-2
Switch4(config-if-range)#sw mo tr
Switch4(config-if-range)#int f0/1
Switch4(config-if)#sw mo acc
Switch4(config-if)#sw acc vlan 20
```

图 16-5　Switch4 配置

Switch5 的接口 f0/23、f0/24、G0/1、G0/2 配置为 trunk 模式,并将 f0/1 接口划分到 VLAN 10 中,如图 16-6 所示。

Switch6 的接口 f0/23、f0/24、G0/1、G0/2 配置为 trunk 模式,并将 f0/1 接口划分到 VLAN 20 中,如图 16-7 所示。

```
Switch5(config)#vlan 10
Switch5(config-vlan)#vlan 20
Switch5(config-vlan)#int r f0/23-24
Switch5(config-if-range)#sw mo tr
Switch5(config-if-range)#int r g0/1-2
Switch5(config-if-range)#sw mo tr
Switch5(config-if-range)#int f0/1
Switch5(config-if)#sw mo acc
Switch5(config-if)#sw acc vlan 10
```

图 16-6　Switch5 配置

```
Switch6(config)#vlan 10
Switch6(config-vlan)#vlan 20
Switch6(config-vlan)#int r f0/23-24
Switch6(config-if-range)#sw mo tr
Switch6(config-if-range)#int r g0/1-2
Switch6(config-if-range)#sw mo tr
Switch6(config-if-range)#int f0/1
Switch6(config-if)#sw mo acc
Switch6(config-if)#sw acc vlan 20
```

图 16-7　Switch6 配置

查看 Switch1 和 Switch2 的生成树信息,可以看到多个连接到核心的线路被堵塞,根被 Switch4 抢走,备用线路被当作主线路使用,如图 16-8 所示。

开始配置生成树。

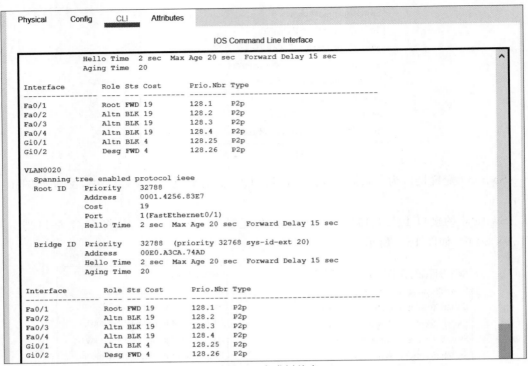

图 16-8　生成树信息

配置 Switch1 为 VLAN 10 的根，配置 Switch1 为 VLAN 20 的次级根，如图 16-9 所示。

```
Switch1(config)#spanning-tree mode pvst
Switch1(config)#spanning-tree vlan 10 root pri
Switch1(config)#spann vlan 20 root sec
```

图 16-9　Switch1 生成树配置

配置 Switch2 为 VLAN 10 的次级根，配置 Switch2 为 VLAN 20 的根，如图 16-10 所示。

```
Switch2(config)#spanning-tree mode pvst
Switch2(config)#spanning-tree vlan 10 root sec
Switch2(config)#spann vlan 20 root pri
```

图 16-10　Switch2 生成树配置

结果验证，可以发现配置完成后 Switch1 上的 VLAN 10 接口全部变为转发状态，如图 16-11 所示。

配置完成后 Switch2 上的 VLAN 20 接口全部变为转发状态，如图 16-12 所示。

从 PC ping 服务器地址，得到回复，说明计算机能够访问服务器，如图 16-13 所示。

```
Switch1
 Physical   Config   CLI   Attributes

                               IOS Command Line Interface

 Gi0/1              Altn BLK 4       128.25    P2p
 Gi0/2              Desg FWD 4       128.26    P2p

 VLAN0010
   Spanning tree enabled protocol ieee
   Root ID    Priority    24586
              Address     00E0.A3CA.74AD
              This bridge is the root
              Hello Time  2 sec  Max Age 20 sec  Forward Delay 15 sec

   Bridge ID  Priority    24586  (priority 24576 sys-id-ext 10)
              Address     00E0.A3CA.74AD
              Hello Time  2 sec  Max Age 20 sec  Forward Delay 15 sec
              Aging Time  20

 Interface        Role Sts Cost      Prio.Nbr Type
 ---------------- ---- --- --------- -------- --------------------------------
 Fa0/1            Desg FWD 19        128.1    P2p
 Fa0/2            Desg FWD 19        128.2    P2p
 Fa0/3            Desg FWD 19        128.3    P2p
 Fa0/4            Desg FWD 19        128.4    P2p
 Gi0/1            Desg FWD 4         128.25   P2p
 Gi0/2            Desg FWD 4         128.26   P2p

 VLAN0020
   Spanning tree enabled protocol ieee
   Root ID    Priority    24596
```

图 16-11　Switch1 接口状态

```
 Physical   Config   CLI   Attributes

                               IOS Command Line Interface

              Address     0002.17D5.6669
              Hello Time  2 sec  Max Age 20 sec  Forward Delay 15 sec
              Aging Time  20

 Interface        Role Sts Cost      Prio.Nbr Type
 ---------------- ---- --- --------- -------- --------------------------------
 Fa0/2            Desg FWD 19        128.2    P2p
 Fa0/3            Altn BLK 19        128.3    P2p
 Fa0/1            Desg FWD 19        128.1    P2p
 Fa0/4            Altn BLK 19        128.4    P2p
 Gi0/1            Root FWD 4         128.25   P2p
 Gi0/2            Altn BLK 4         128.26   P2p

 VLAN0020
   Spanning tree enabled protocol ieee
   Root ID    Priority    24596
              Address     0002.17D5.6669
              This bridge is the root
              Hello Time  2 sec  Max Age 20 sec  Forward Delay 15 sec

   Bridge ID  Priority    24596  (priority 24576 sys-id-ext 20)
              Address     0002.17D5.6669
              Hello Time  2 sec  Max Age 20 sec  Forward Delay 15 sec
              Aging Time  20

 Interface        Role Sts Cost      Prio.Nbr Type
 ---------------- ---- --- --------- -------- --------------------------------
 Fa0/2            Desg FWD 19        128.2    P2p
 Fa0/3            Desg FWD 19        128.3    P2p
 Fa0/1            Desg FWD 19        128.1    P2p
 Fa0/4            Desg FWD 19        128.4    P2p
 Gi0/1            Desg FWD 4         128.25   P2p
 Gi0/2            Desg FWD 4         128.26   P2p
```

图 16-12　Switch2 接口状态

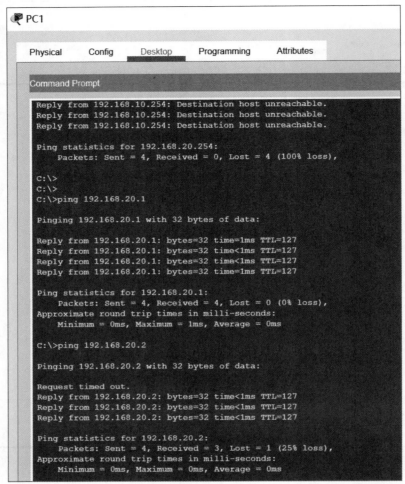

图 16-13　PC 验证

五、实验总结

通过这个实验,学生可以学习到如何在冗余的环境下避免产生广播风暴等问题,验证了生成树协议(STP)的原理。学到冗余线路上端口转发状态的切换方法,以此保证网络的正常通信。学习了为了降低访问服务器速度所产生的影响,避免任意接入交换机随意访问核心交换机和服务器,一般把核心交换机设置为生成树主根以及这样的设置对网络结构的影响。

六、实验思考题

(1) 请思考生成树协议是通过什么样的方式解决环路问题的。
(2) 当链路出现问题时应该如何排错?

实验十七 OSPF 协议的配置与验证

最短路径优先(open shortest path first,OSPF)协议是一种典型的链路状态路由协议,通常在同一路由域里使用。此处,路由域是指一个自治系统(AS),它是指一组通过统一路径,即路由策略或者是路由协议相互交换路由信息的网络。在这个 AS 中,所有的 OSPF 路由器只维护一个相同的描述 AS 结构的数据库,该数据库中存放的是路由域中相应链路的状态信息,OSPF 路由器正是通过这个数据库计算出其 OSPF 路由表的。

OSPF 使用 IP 封装,协议号码为 89,更新方式为组播更新 224.0.0.5(All SPF router)和 224.0.0.6(DR/BDR),触发式更新,每 10s 发送一次 Hello 包,40s 未收到视为邻居故障。

OSPF 的特点如下。

(1) 把自治系统在逻辑意义上分成一个或者多个区域(减少 SPF 计算量)。
(2) OSPF 可以以 LSA(link state advertisement)的形式发布路由。
(3) 通过区域内设备间报文的交流,达到路由信息交互。
(4) OSPF 的报文封装在 IP 报文内,通常采用单播或组播的形式发送。
(5) 允许等开销负载均衡。
(6) 支持路由认证。
(7) 支持外部路由的注入,同时能够识别外部路由。
(8) 支持无类网络。

一、实验目的

(1) 理解 OSPF 协议的基本原理。
(2) 掌握 OSPF 的常用配置方法。
(3) 熟悉路由器的路由表。

二、实验环境

Cisco Packet Tracer 6.0 模拟器(路由器、交换机、直通线、交叉线、计算机)。

三、实验内容及步骤

(1) 搭建实验拓扑。
(2) 配置接口 IP 信息。

(3) 配置 OSPF 协议，创建 2 个区域。
(4) 部分链路配置 OSPF 加密。
(5) 查看 OSPF 邻居信息。
(6) 网络测试。

四、实验过程

1. 构建实验拓扑

根据实验内容，构建实验拓扑，如图 17-1 所示。

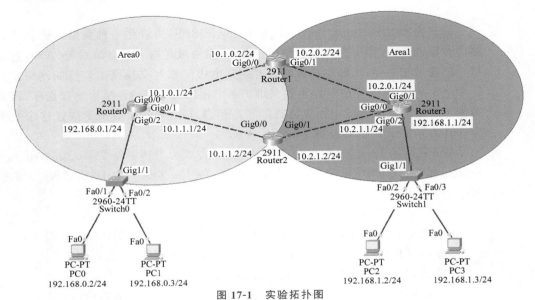

图 17-1 实验拓扑图

2. 地址规划

进行合理的 IP 地址规划，如表 17-1 所示。

表 17-1 IP 地址规划表

设 备	接 口	IP 地址
Router0	GE0/1	10.1.1.1/24
Router0	GE0/2	192.168.0.1/24
Router1	GE0/0	10.1.0.2/24
Router1	GE0/1	10.2.0.2/24
Router2	GE0/0	10.1.1.2/24
Router2	GE0/1	10.2.1.2/24
Router3	GE0/0	10.2.0.1/24
Router3	GE0/1	10.2.1.1/24
Router3	GE0/2	192.168.1.1/24

续表

设　　备	接　　口	IP 地 址
PC0	Fa0	192.168.0.2/24
PC1	Fa0	192.168.0.3/24
PC2	Fa0	192.168.1.2/24
PC3	Fa0	192.168.1.3/24

3. 设备配置

（1）Router1 接口配置，如图 17-2 所示。

```
Router>en
Router#conf t
Enter configuration commands, one per line.  End with CNTL/Z.
Router(config)#int g0/2
Router(config-if)#ip add 192.168.0.1 255.255.255.0
Router(config-if)#no shut

Router(config-if)#
%LINK-5-CHANGED: Interface GigabitEthernet0/2, changed state to up

%LINEPROTO-5-UPDOWN: Line protocol on Interface GigabitEthernet0/2, changed sta
e to up
int g0/1
Router(config-if)#ip add 10.1.1.1 255.255.255.0
Router(config-if)#no shut

Router(config-if)#
%LINK-5-CHANGED: Interface GigabitEthernet0/1, changed state to up
int g0/0
Router(config-if)#ip add 10.1.0.1 255.255.255.0
Router(config-if)#no shut

Router(config-if)#
%LINK-5-CHANGED: Interface GigabitEthernet0/0, changed state to up
```

图 17-2　Router1 接口配置

（2）Router1 接口 OSPF 加密配置，如图 17-3 所示。

```
R1(config)#router ospf 1
R1(config-router)# log-adjacency-changes
R1(config-router)# network 192.168.0.0 0.0.0.255 area 0
R1(config-router)# network 10.1.0.0 0.0.0.255 area 0
R1(config-router)# network 10.1.1.0 0.0.0.255 area 0
R1(config-router)#
R1(config-router)#
R1(config-router)#
R1(config-router)#int g0/0
R1(config-if)#ip ospf authentication message-digest
R1(config-if)# ip ospf authentication-key cisco
```

图 17-3　Router1 路由器接口 OSPF 加密配置

（3）Router2 路由器接口配置，如图 17-4 所示。

按照同样的方式对 Router2 进行 OSPF 加密配置。

（4）Router3 接口配置，如图 17-5 所示。

（5）按照同样的方式，参照表 17-1 的地址规划方案，完成对 Router0 的配置。

```
Router(config)#hostname Router2
Router2(config)#
Router2(config)#
Router2(config)#router ospf 1
Router2(config-router)#rou
Router2(config-router)#router-id 2.2.2.2
Router2(config-router)#network 10.1.1.0 0.0.0.255 area 0
Router2(config-router)#network 10.2.1.0 0.0.0.255 area 1
Router2(config-router)#no
00:55:45: %OSPF-5-ADJCHG: Process 1, Nbr 192.168.0.1 on Gi
LOADING to FULL, Loading Done
% Ambiguous command: "n"
```

图 17-4 Router2 接口配置

```
Router>enable
Router#configure terminal
Enter configuration commands, one per line.  End with CNTl
Router(config)#hostname Router3
Router3(config)#int g0/1
Router3(config-if)#ip add 10.2.0.1 255.255.255.0
Router3(config-if)#no shut

Router3(config-if)#
%LINK-5-CHANGED: Interface GigabitEthernet0/1, changed sta

%LINEPROTO-5-UPDOWN: Line protocol on Interface GigabitEth
e to up
int g0/0
Router3(config-if)#ip add 10.2.1.1 255.255.255.0
Router3(config-if)#no shut

Router3(config-if)#
%LINK-5-CHANGED: Interface GigabitEthernet0/0, changed sta

%LINEPROTO-5-UPDOWN: Line protocol on Interface GigabitEth
e to up
int g0/2
Router3(config-if)#ip add 192.168.1.1 255.255.255.0
Router3(config-if)#no shut
```

图 17-5 Router3 接口配置

4. 结果验证

(1) 查看路由器 Router0 的 OSPF 状态,主要验证邻居建立状态是否为 full,以及是否接收到 OIA 路由,如图 17-6 和图 17-7 所示。

```
Router0#sh ip ospf neighbor

Neighbor ID     Pri   State        Dead Time   Address      Interface
2.2.2.2          1    FULL/BDR     00:00:30    10.1.1.2     GigabitEtherne
t0/1
10.2.0.2         1    FULL/BDR     00:00:33    10.1.0.2     GigabitEtherne
t0/0
Router0#sh ip ospf neighbor

Neighbor ID     Pri   State        Dead Time   Address      Interface
2.2.2.2          1    FULL/BDR     00:00:35    10.1.1.2     GigabitEtherne
t0/1
10.2.0.2         1    FULL/BDR     00:00:38    10.1.0.2     GigabitEtherne
t0/0
```

图 17-6 查看 Router0 的 OSPF 状态

(2) 查看路由器 Router1 的 OSPF 状态,主要验证邻居建立状态是否为 full,以及是否接收到 OIA 路由,如图 17-8 所示。

(3) 查看路由器 Router2 的 OSPF 状态,主要验证邻居建立状态是否为 full,以及是否接收到 OIA 路由,如图 17-9 所示。

```
Gateway of last resort is not set

     10.0.0.0/8 is variably subnetted, 6 subnets, 2 masks
C       10.1.0.0/24 is directly connected, GigabitEthernet0/0
L       10.1.0.1/32 is directly connected, GigabitEthernet0/0
C       10.1.1.0/24 is directly connected, GigabitEthernet0/1
L       10.1.1.1/32 is directly connected, GigabitEthernet0/1
O IA    10.2.0.0/24 [110/2] via 10.1.0.2, 00:04:53, GigabitEthernet0/0
O IA    10.2.1.0/24 [110/2] via 10.1.1.2, 00:04:43, GigabitEthernet0/1
     192.168.0.0/24 is variably subnetted, 2 subnets, 2 masks
C       192.168.0.0/24 is directly connected, GigabitEthernet0/2
L       192.168.0.1/32 is directly connected, GigabitEthernet0/2
O IA 192.168.1.0/24 [110/3] via 10.1.0.2, 00:03:12, GigabitEthernet0/0
                    [110/3] via 10.1.1.2, 00:03:12, GigabitEthernet0/1
```

图 17-7　查看 Router0 的 OSPF 状态（续）

```
Router1#sh ip ospf neighbor

Neighbor ID     Pri   State           Dead Time   Address         Interface
192.168.0.1       1   FULL/DR         00:00:30    10.1.0.1        GigabitEthernet0/0
3.3.3.3           1   FULL/BDR        00:00:36    10.2.0.1        GigabitEthernet0/1
Router1#sh ip rou
Router1#sh ip route
Codes: L - local, C - connected, S - static, R - RIP, M - mobile, B - BGP
       D - EIGRP, EX - EIGRP external, O - OSPF, IA - OSPF inter area
       N1 - OSPF NSSA external type 1, N2 - OSPF NSSA external type 2
       E1 - OSPF external type 1, E2 - OSPF external type 2, E - EGP
       i - IS-IS, L1 - IS-IS level-1, L2 - IS-IS level-2, ia - IS-IS inter area
       * - candidate default, U - per-user static route, o - ODR
       P - periodic downloaded static route

Gateway of last resort is not set

     10.0.0.0/8 is variably subnetted, 6 subnets, 2 masks
C       10.1.0.0/24 is directly connected, GigabitEthernet0/0
L       10.1.0.2/32 is directly connected, GigabitEthernet0/0
O       10.1.1.0/24 [110/2] via 10.1.0.1, 00:06:57, GigabitEthernet0/0
C       10.2.0.0/24 is directly connected, GigabitEthernet0/1
L       10.2.0.2/32 is directly connected, GigabitEthernet0/1
O       10.2.1.0/24 [110/2] via 10.2.0.1, 00:04:18, GigabitEthernet0/1
O    192.168.0.0/24 [110/2] via 10.1.0.1, 00:10:24, GigabitEthernet0/0
O    192.168.1.0/24 [110/2] via 10.2.0.1, 00:04:18, GigabitEthernet0/1
```

图 17-8　查看 Router1 的 OSPF 状态

```
Router2#sh ip ospf neighbor

Neighbor ID     Pri   State           Dead Time   Address         Interface
192.168.0.1       1   FULL/DR         00:00:39    10.1.1.1        GigabitEthernet0/0
3.3.3.3           1   FULL/BDR        00:00:36    10.2.1.1        GigabitEthernet0/1
Router2#sh ip rou
Router2#sh ip route
Codes: L - local, C - connected, S - static, R - RIP, M - mobile, B - BGP
       D - EIGRP, EX - EIGRP external, O - OSPF, IA - OSPF inter area
       N1 - OSPF NSSA external type 1, N2 - OSPF NSSA external type 2
       E1 - OSPF external type 1, E2 - OSPF external type 2, E - EGP
       i - IS-IS, L1 - IS-IS level-1, L2 - IS-IS level-2, ia - IS-IS inter area
       * - candidate default, U - per-user static route, o - ODR
       P - periodic downloaded static route

Gateway of last resort is not set

     10.0.0.0/8 is variably subnetted, 6 subnets, 2 masks
O       10.1.0.0/24 [110/2] via 10.1.1.1, 00:08:25, GigabitEthernet0/0
C       10.1.1.0/24 is directly connected, GigabitEthernet0/0
L       10.1.1.2/32 is directly connected, GigabitEthernet0/0
O       10.2.0.0/24 [110/2] via 10.2.1.1, 00:05:46, GigabitEthernet0/1
C       10.2.1.0/24 is directly connected, GigabitEthernet0/1
L       10.2.1.2/32 is directly connected, GigabitEthernet0/1
O    192.168.0.0/24 [110/2] via 10.1.1.1, 00:08:25, GigabitEthernet0/0
O    192.168.1.0/24 [110/2] via 10.2.1.1, 00:05:46, GigabitEthernet0/1
```

图 17-9　查看 Router2 的 OSPF 状态

(4) 查看路由器 Router3 的 OSPF 状态,主要验证邻居建立状态是否为 full,以及是否接收到 OIA 路由,如图 17-10 所示。

```
Router3#sh ip ospf neighbor
Neighbor ID     Pri   State         Dead Time    Address        Interface
10.2.0.2        1     FULL/DR       00:00:32     10.2.0.2       GigabitEtherne
t0/1
2.2.2.2         1     FULL/DR       00:00:31     10.2.1.2       GigabitEtherne
t0/0
Router3#sh ip rou
Router3#sh ip route
Codes: L - local, C - connected, S - static, R - RIP, M - mobile, B - BGP
       D - EIGRP, EX - EIGRP external, O - OSPF, IA - OSPF inter area
       N1 - OSPF NSSA external type 1, N2 - OSPF NSSA external type 2
       E1 - OSPF external type 1, E2 - OSPF external type 2, E - EGP
       i - IS-IS, L1 - IS-IS level-1, L2 - IS-IS level-2, ia - IS-IS inter area
       * - candidate default, U - per-user static route, o - ODR
       P - periodic downloaded static route

Gateway of last resort is not set

     10.0.0.0/8 is variably subnetted, 6 subnets, 2 masks
O IA    10.1.0.0/24 [110/2] via 10.2.0.2, 00:06:35, GigabitEthernet0/1
O IA    10.1.1.0/24 [110/2] via 10.2.1.2, 00:06:35, GigabitEthernet0/0
C       10.2.0.0/24 is directly connected, GigabitEthernet0/1
L       10.2.0.1/32 is directly connected, GigabitEthernet0/1
C       10.2.1.0/24 is directly connected, GigabitEthernet0/0
L       10.2.1.1/32 is directly connected, GigabitEthernet0/0
O IA 192.168.0.0/24 [110/3] via 10.2.0.2, 00:06:35, GigabitEthernet0/1
                    [110/3] via 10.2.1.2, 00:06:35, GigabitEthernet0/0
     192.168.1.0/24 is variably subnetted, 2 subnets, 2 masks
C       192.168.1.0/24 is directly connected, GigabitEthernet0/2
L       192.168.1.1/32 is directly connected, GigabitEthernet0/2
```

图 17-10 查看 Router3 的 OSPF 状态

(5) 测试 PC0 分别到 PC2、PC3 的连通性,结果如图 17-11 所示。

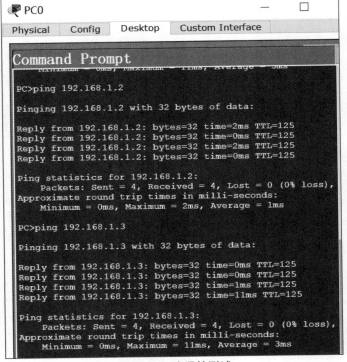

图 17-11 连通性测试

五、实验总结

本实验可以让学生进一步了解 OSPF 技术在网络中的应用,学会搭建 OSPF 环境,实现通过 OSPF 完成内部网络的通信。本实验还验证了 OSPF 在不同区域连接时的现象,针对特殊线路做 OSPF 加密时的表现。

六、实验思考题

(1) OSPF 建邻关系复杂,受到多种因素影响,那么 OSPF 不能建邻的原因是什么?

(2) 路由优先级在协议不同时,如何进行路由选路?

(3) OSPF 在选举 DR/BDR 时是如何选举的?

BGP 的配置与验证

实验十八

BGP(border gateway protocol,边界网关协议)是一种外部路由协议,通常是在 AS 之前传递各个路由信息的。与一般的 IGP 不一样,BGP 主要用来实现路由控制。BGP 是增强型的距离矢量路由协议,依靠准确的路由更新机制、多类型的 Metric 测量方法,从设计上来避免产生环路(AS-PATH)的问题,并且路由也可以附带一些属性信息、CIDR 汇总以及路由策略。

BGP 共发展了 4 个版本:RFC1105(BGPV1)、RFC1163(BGPV2)、RFC1267(BGPV3),以及当前使用的 RFC4271/RFC1771(BGPV4)。BGPV4 目前正在快速成为边界路由协议的标杆。

BGP 基于 TCP 的端口号为 179,通过触发式更新,只发送自身增量路由。周期性发送 Keepalive 报文,用来监测 TCP 是否存活。BGP 建邻依赖于基础路由协议,需建立 TCP 连接。

BGP 对等体的建立、更新和删除等交互过程主要有以下 5 种报文、6 种状态机和 5 个原则。报文包含 Open 报文、Update 报文、Notification 报文、Keepalive 报文、Route-refresh 报文。BGP 状态机包含空闲时刻(Idle)、连接时刻(Connect)、活跃时刻(Active)、Open 报文成功送达(OpenSent)、Open 报文成功确认(OpenConfirm)和连接成功建立(Established)。交互的原则包括当 BGP 路由由 IBGP 对等体获得时,BGP 设备只发布给它的 EBGP 对等体;从 EBGP 对等体获得的 BGP 路由,BGP 设备只会发布给其所有 EBGP 和 IBGP 对等体;若有多条有效路由到达同一目的地址时,BGP 设备只会将最优路由发给对等体;当路由更新时,BGP 设备只发送已经更新了的 BGP 路由;所有被对等体发送的路由,BGP 设备全部会接收。

一、实验目的

(1) 理解 BGP 的基本原理。
(2) 掌握 BGP 的常用配置方法。

二、实验环境

eNSP V100R003C00SPC100 模拟器(路由器、交换机、直通线、计算机)。

三、实验内容及步骤

(1) 搭建实验拓扑。

（2）配置接口 IP 信息。
（3）配置 BGP。
（4）查看 BGP 邻居信息。
（5）网络测试。

四、实验过程

1. 实验拓扑

实验拓扑如图 18-1 所示。

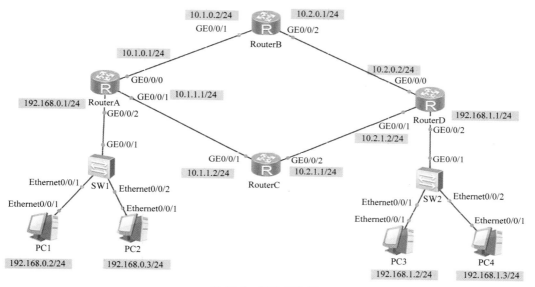

图 18-1 实验拓扑图

2. 地址规划

IP 地址规划如表 18-1 所示。

表 18-1 IP 地址规划表

序　号	设　备	接　口	IP 地址
1	RouterA	GE0/0/1	10.1.1.1/24
2	RouterA	GE0/0/2	192.168.0.1/24
3	RouterB	GE0/0/1	10.1.0.2/24
4	RouterB	GE0/0/2	10.2.0.1/24
5	RouterC	GE0/0/1	10.1.1.2/24
6	RouterC	GE0/0/2	10.2.1.1/24
7	RouterD	GE0/0/0	10.2.0.2/24
8	RouterD	GE0/0/1	10.2.1.2/24
9	RouterD	GE0/0/2	192.168.1.1/24

序 号	设 备	接 口	IP 地 址
10	PC1	Eth0/0/1	192.168.0.2/24
11	PC2	Eth0/0/1	192.168.0.3/24
12	PC3	Eth0/0/1	192.168.1.2/24
13	PC4	Eth0/0/1	192.168.1.3/24

3. 实验配置

(1) RouterA 的配置如图 18-2 所示。

```
<RouterA>sys
[RouterA]interface GigabitEthernet0/0/0              //配置接口 G0/0/0 的 IP 地址
[RouterA-GigabitEthernet0/0/0]ip address 10.1.0.1 255.255.255.0
[RouterA-GigabitEthernet0/0/0]quit
[RouterA]interface GigabitEthernet0/0/1              //配置接口 G0/0/1 的 IP 地址
[RouterA-GigabitEthernet0/0/1]ip address 10.1.1.1 255.255.255.0
[RouterA-GigabitEthernet0/0/1]quit
[RouterA]interface GigabitEthernet0/0/2              //配置接口 G0/0/2 的 IP 地址
[RouterA-GigabitEthernet0/0/2]ip address 192.168.0.1 255.255.255.0
[RouterA-GigabitEthernet0/0/2]quit
[RouterA]bgp 65001                                   //启用 BGP 并配置 Router ID
[RouterA-bgp]peer 10.1.0.2 as-number 65002           //配置 BGP 邻居信息
[RouterA-bgp]peer 10.1.1.2 as-number 65003
[RouterA-bgp]network 10.1.0.0 24                     //配置广播地址
[RouterA-bgp]network 10.1.1.0 24
[RouterA-bgp]network 192.168.0.0 24
<RouterA>save                                        //保存配置
The current configuration will be written to the device.
Are you sure to continue? (y/n)[n]:y
It will take several minutes to save configuration file, please wait........
Configuration file had been saved successfully
Note: The configuration file will take effect after being activated
```

图 18-2 RouterA 配置

(2) RouterB 的配置如图 18-3 所示。

(3) RouterC 的配置如图 18-4 所示。

(4) RouterD 的配置如图 18-5 所示。

4. 结果验证

(1) 查看 RouterA BGP 路由信息,结果如图 18-6 所示。

(2) 查看 RouterB BGP 路由信息,结果如图 18-7 所示。

(3) 查看 RouterC BGP 路由信息,结果如图 18-8 所示。

(4) 查看 RouterD BGP 路由信息,结果如图 18-9 所示。

```
<RouterB>sys
[RouterB]interface GigabitEthernet0/0/1        //配置接口 G0/0/1 的 IP 地址
[RouterB-GigabitEthernet0/0/1]ip address 10.1.0.2 255.255.255.0
[RouterB]interface GigabitEthernet0/0/2        //配置接口 G0/0/2 的 IP 地址
[RouterB-GigabitEthernet0/0/2]ip address 10.2.0.1 255.255.255.0
[RouterB]bgp 65002                             //启用 BGP 并配置 Router ID
[RouterB-bgp]router-id 2.2.2.2
[RouterB-bgp]peer 10.1.0.1 as-number 65001     //配置 BGP 邻居信息
[RouterB-bgp]peer 10.2.0.2 as-number 65004
[RouterB-bgp]network 10.1.0.0 24               //配置广播地址
[RouterB-bgp]network 10.2.0.0 24
[RouterB-bgp]quit
<RouterB>save                                  //保存配置
The current configuration will be written to the device.
Are you sure to continue? (y/n)[n]:y
It will take several minutes to save configuration file, please wait.......
Configuration file had been saved successfully
Note: The configuration file will take effect after being activated
<RouterB>
```

图 18-3 RouterB 配置

```
<RouterC>sys
[RouterC]interface GigabitEthernet0/0/1        //配置接口 G0/0/1 的 IP 地址
[RouterC-GigabitEthernet0/0/1]ip address 10.1.1.2 255.255.255.0
[RouterC]interface GigabitEthernet0/0/2        //配置接口 G0/0/2 的 IP 地址
[RouterC-GigabitEthernet0/0/2]ip address 10.2.1.1 255.255.255.0
[RouterC]bgp 65003                             //启用 BGP 并配置 Router ID
[RouterC-bgp]router-id 3.3.3.3
[RouterC-bgp]peer 10.1.1.1 as-number 65001     //配置 BGP 邻居信息
[RouterC-bgp]peer 10.2.1.2 as-number 65004
[RouterC-bgp]network 10.1.1.0 24               //配置广播地址
[RouterC-bgp]network 10.2.1.0 24
<RouterC>save                                  //保存配置
The current configuration will be written to the device.
Are you sure to continue? (y/n)[n]:y
It will take several minutes to save configuration file, please wait.......
Configuration file had been saved successfully
Note: The configuration file will take effect after being activated
<RouterC>
```

图 18-4 RouterC 配置

```
< RouterD >sys
[RouterD]interface GigabitEthernet0/0/0          //配置接口 G0/0/0 的 IP 地址
[RouterD-GigabitEthernet0/0/0]ip address 10.2.0.2 255.255.255.0
[RouterD]interface GigabitEthernet0/0/1          //配置接口 G0/0/1 的 IP 地址
[RouterD-GigabitEthernet0/0/1]ip address 10.2.1.2 255.255.255.0
[RouterD]interface GigabitEthernet0/0/2          //配置接口 G0/0/2 的 IP 地址
[RouterD-GigabitEthernet0/0/2]ip address 192.168.1.1 255.255.255.0
[Huawei]bgp 65004                                //启用 BGP 并配置 Router ID
[Huawei-bgp]router-id 4.4.4.4
[Huawei-bgp]peer 10.2.0.1 as-number 65002        //配置 BGP 邻居信息
[Huawei-bgp]peer 10.2.1.1 as-number 65003
[Huawei-bgp]network 10.2.0.0 24                  //配置广播地址
[Huawei-bgp]network 10.2.1.0 24
[Huawei-bgp]network 192.168.1.0 24
<Huawei>save                                     //保存配置
The current configuration will be written to the device.
Are you sure to continue? (y/n)[n]:y
It will take several minutes to save configuration file, please wait.......
Configuration file had been saved successfully
Note: The configuration file will take effect after being activated
<Huawei>
```

图 18-5 RouterD 配置

```
RouterA
<RouterA>disp bgp peer
 BGP local router ID : 1.1.1.1
 Local AS number : 65001
 Total number of peers : 2        Peers in established state : 2

  Peer          V       AS    MsgRcvd  MsgSent  OutQ  Up/Down     State  Pre
fRcv
  10.1.0.2      4     65002      24       26      0  00:15:42  Established
4
  10.1.1.2      4     65003      21       23      0  00:12:51  Established
4
```

图 18-6 RouterA 验证结果

```
RouterB
<RouterB>disp bgp peer
 BGP local router ID : 2.2.2.2
 Local AS number : 65002
 Total number of peers : 2        Peers in established state : 2

  Peer          V       AS    MsgRcvd  MsgSent  OutQ  Up/Down     State  Pre
fRcv
  10.1.0.1      4     65001      27       26      0  00:17:08  Established
4
  10.2.0.2      4     65004      16       17      0  00:07:21  Established
4
```

图 18-7 RouterB 验证结果

```
RouterC
<RouterC>disp bgp peer
 BGP local router ID : 10.1.1.2
 Local AS number : 65003
 Total number of peers : 2        Peers in established state : 2

    Peer          V      AS    MsgRcvd  MsgSent   OutQ  Up/Down      State  Pre
fRcv
    10.1.1.1      4    65001     24       23       0  00:14:53   Established
     5
    10.2.1.2      4    65004     16       18       0  00:07:37   Established
     5
```

图 18-8　RouterC 验证结果

```
RouterD
<Huawei>disp bgp peer
 BGP local router ID : 4.4.4.4
 Local AS number : 65004
 Total number of peers : 2        Peers in established state : 2

    Peer          V      AS    MsgRcvd  MsgSent   OutQ  Up/Down      State  Pre
fRcv
    10.2.0.1      4    65002     17       17       0  00:08:44   Established
     4
    10.2.1.1      4    65003     17       17       0  00:08:23   Established
     4
```

图 18-9　RouterD 验证结果

（5）测试 PC2 到 PC3 和 PC4 的网络连通是否正常，验证路由的准确性，在此时一定要注意计算机的网关配置。

五、实验总结

本实验能够让学生进一步了解 BGP 技术在网络中的应用，学会搭建 BGP 环境，实现通过 BGP 完成内部网络的通信。

六、实验思考题

（1）请说出 BGP 无法建邻的原因。
（2）如果 BGP 邻居关系一直停留在 idle 状态，请陈述原因。
（3）如果 BGP 邻居关系一直停留在 active 状态，请陈述原因。

实验十九

VRRP 的配置与验证

VRRP(virtual router redundancy protocol,虚拟路由冗余协议)可以把局域网内一组路由器(包括一台主路由器和多台备份路由器)组织成一台虚拟路由器,称作备份组。当 master 设备出现故障时,虚拟路由将启用备份路由器,从而实现网络通信冗余,保证通信正常。

VRRP 包含 3 种状态机:原始状态(initialize)、活跃状态(master)、备份状态(backup)。其中,只有处于 master 状态的设备才具备转发能力,才可以转发那些发送到虚拟 IP 地址的报文。

VRRP 的工作过程如下。

(1) 先比较 VRRP 备份组中设备的优先级,然后选择主设备。主设备发送免费的 ARP 报文,并将虚拟 MAC 地址发给其连接的设备或者主机,由它们进行报文转发。

(2) master 设备周期性向备份组内所有 backup 设备发送 VRRP 通告报文,宣布配置信息和状态。

(3) 当主设备死机,VRRP 备份组中的 backup 设备会依据优先级重新选主设备。

(4) 切换 VRRP 备份组状态时,新的主设备会立即发送一条免费的 ARP 信息,该信息中包含虚拟路由器的虚拟 MAC 地址和虚拟 IP 地址信息,刷新与其连接的主机或设备的 MAC 条目,从而将用户流量引导到新的主设备,整个过程对用户来说是完全透明的。

(5) 原主设备恢复正常时,若设备为 IP 地址所属者(优先级为 255),则直接切换到 master 状态。如若设备优先级低于 255,会最先切换成 backup 状态,并且优先级变成故障前设置的优先级。

(6) 当 backup 设备的优先级高于 master 设备时,由 backup 设备的工作方式决定要不要重新选举 master。

一、实验目的

(1) 理解 VRRP 的基本原理。
(2) 掌握 VRRP 的常用配置方法。
(3) 熟悉路由器的路由表。

二、实验环境

eNSP V100R003C00SPC100 模拟器(路由器、交换机、直通线、计算机)。

三、实验内容及步骤

(1)搭建实验拓扑。
(2)配置接口 IP 信息。
(3)配置 VRRP。
(4)查看 VRRP 邻居信息。
(5)网络测试。

四、实验过程

1. 实验拓扑

根据实验内容,设计网络拓扑,如图 19-1 所示。

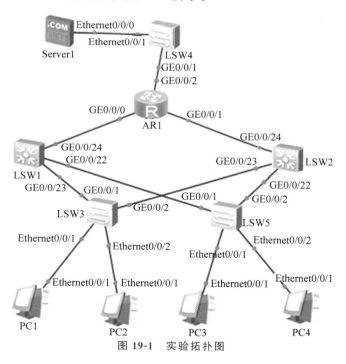

图 19-1 实验拓扑图

2. 地址规划

参照拓扑图,合理规划 IP 地址,如表 19-1 所示。

表 19-1 IP 地址规划表

1	PC1	FastEthernet0	192.168.10.1	255.255.255.0
2	PC2	FastEthernet0	192.168.20.1	255.255.255.0
3	PC3	FastEthernet0	192.168.30.1	255.255.255.0

				续表
4	PC4	FastEthernet0	192.168.40.1	255.255.255.0
5	Server0	FastEthernet0	192.168.50.1	255.255.255.0

3. 实验配置

(1) 更改接口模式,配置 VLAN 信息,开通 trunk 链路,如图 19-2 所示。

```
[Switch1-vlan200]int vlan 200
[Switch1-Vlanif200]ip add 192.168.200.1 255.255.255.252
[Switch1-Vlanif200]int g0/0/24
[Switch1-GigabitEthernet0/0/24]
[Switch1-GigabitEthernet0/0/24]port link-type access
[Switch1-GigabitEthernet0/0/24]port default vlan 200
[Switch1-port-group]port link-type trunk
[Switch1-GigabitEthernet0/0/22]port link-type trunk
[Switch1-GigabitEthernet0/0/23]port link-type trunk
[Switch1]int g0/0/24
[Switch1-GigabitEthernet0/0/24]port link-type access
[Switch1-GigabitEthernet0/0/24]port default vlan 100
```

图 19-2　Switch1 配置

(2) 配置端口组,放行所有 VLAN。在 Switch2 上配置权限,允许所有 VLAN 通过,如图 19-3 所示。

```
[Switch2]port-group group-member GigabitEthernet 0/0/22 to GigabitEthernet 0/0/23
[Switch2-port-group]port link-type trunk
[Switch2-GigabitEthernet0/0/22]port link-type trunk
[Switch2-GigabitEthernet0/0/23]port link-type trunk
[Switch2-GigabitEthernet0/0/22]port trun all vlan all
[Switch2-GigabitEthernet0/0/23]port trun all vlan all
[Switch2-port-group]
[Switch2-vlan200]int vlan 200
[Switch2-Vlanif200]ip add 192.168.200.1 255.255.255.252
[Switch2-Vlanif200]int g0/0/24
[Switch2-GigabitEthernet0/0/24]
[Switch2-GigabitEthernet0/0/24]port link-type access
[Switch2-GigabitEthernet0/0/24]port default vlan 200
```

图 19-3　Switch2 配置

(3) 配置默认 VLAN,放行所有 VLAN。在 Switch3 上配置权限,允许所有 VLAN 通过,如图 19-4 所示。

(4) 在 Switch5 上配置相应 VLAN,并允许所有 VLAN 通过,如图 19-5 所示。

(5) Router1 接口配置。进入路由器 Router1 的 g0/0、g0/1 和 g0/2 接口,开启端口,并配置端口的 IP 地址,如图 19-6 所示。

```
[Switch3-vlan40]int e0/0/1
[Switch3-Ethernet0/0/1]port link-ty acc
[Switch3-Ethernet0/0/1]port def vlan 10
[Switch3-Ethernet0/0/1]int e0/0/2
[Switch3-Ethernet0/0/2]port link-ty acc
[Switch3-Ethernet0/0/2]port de vlan 20
[Switch3]port-group group-member GigabitEthernet 0/0/1 to GigabitEthernet 0/0/2
[Switch3-port-group]port trunk all vlan all
[Switch3-GigabitEthernet0/0/1]port trunk all vlan all
[Switch3-GigabitEthernet0/0/2]port trunk all vlan all
```

图 19-4　Switch3 配置

```
[Switch5-vlan40]int e0/0/1
[Switch5-Ethernet0/0/1]port link-ty acc
[Switch5-Ethernet0/0/1]port def vlan 30
[Switch5-Ethernet0/0/1]int e0/0/2
[Switch5-Ethernet0/0/2]port link-ty acc
[Switch5-Ethernet0/0/2]port de vlan 40
[Switch5-Ethernet0/0/2]
[Switch5]int g0/0/1
[Switch5-GigabitEthernet0/0/1]port link-ty trunk
[Switch5-GigabitEthernet0/0/1]port trunk all vlan all
[Switch5-GigabitEthernet0/0/1]int g0/0/2
[Switch5-GigabitEthernet0/0/2]port link-ty trunk
[Switch5-GigabitEthernet0/0/2]port trunk all vlan all
```

图 19-5　Switch5 配置

```
[R1]int g0/0/0
[R1-GigabitEthernet0/0/0]ip add 192.168.100.2 255.255.255.252
[R1-GigabitEthernet0/0/0]int g0/0/1
[R1-GigabitEthernet0/0/1]ip add 192.168.200.2 255.255.255.252
[R1-GigabitEthernet0/0/1]int g0/0/2
[R1-GigabitEthernet0/0/2]ip add 192.168.50.254 255.255.255.0
```

图 19-6　Router1 的配置

（6）在路由器 Router1 上启动 RIP,宣告直连网络,如图 19-7 所示。
（7）Switch1 宣告本机网段,如图 19-8 所示。

```
[R1]rip 1
[R1-rip-1]network 192.168.100.0
[R1-rip-1]network 192.168.200.0
[R1-rip-1]network 192.168.50.0
```

图 19-7　Router1 网段宣告

```
[Switch1]rip 1
[Switch1-rip-1]net 192.168.100.0
[Switch1-rip-1]net 192.168.10.0
[Switch1-rip-1]net 192.168.20.0
[Switch1-rip-1]net 192.168.30.0
[Switch1-rip-1]net 192.168.40.0
```

图 19-8　Switch1 网段宣告

（8）Switch2 宣告本机网段，如图 19-9 所示。

```
[Switch2]rip 1
[Switch2-rip-1]network 192.168.10.0
[Switch2-rip-1]network 192.168.20.0
[Switch2-rip-1]network 192.168.30.0
[Switch2-rip-1]network 192.168.40.0
[Switch2-rip-1]network 192.168.200.0
```

图 19-9　Switch2 网段宣告

（9）查看路由表，如图 19-10 所示。

```
Route Flags: R - relay, D - download to fib

Routing Tables: Public
         Destinations : 13        Routes : 13

Destination/Mask    Proto   Pre  Cost      Flags NextHop         Interface
       127.0.0.0/8  Direct  0    0         D     127.0.0.1       InLoopBack0
      127.0.0.1/32  Direct  0    0         D     127.0.0.1       InLoopBack0
    192.168.10.0/24 RIP     100  2         D     192.168.200.2   Vlanif200
    192.168.20.0/24 RIP     100  2         D     192.168.200.2   Vlanif200
    192.168.30.0/24 Direct  0    0         D     192.168.30.253  Vlanif30
  192.168.30.253/32 Direct  0    0         D     127.0.0.1       Vlanif30
    192.168.40.0/24 Direct  0    0         D     192.168.40.253  Vlanif40
  192.168.40.253/32 Direct  0    0         D     127.0.0.1       Vlanif40
    192.168.50.0/24 RIP     100  1         D     192.168.200.2   Vlanif200
   192.168.100.0/24 RIP     100  1         D     192.168.200.2   Vlanif200
   192.168.200.0/24 RIP     100  1         D     192.168.200.2   Vlanif200
   192.168.200.0/30 Direct  0    0         D     192.168.200.1   Vlanif200
   192.168.200.1/32 Direct  0    0         D     127.0.0.1       Vlanif200
[Switch2]
```

图 19-10　路由表信息

五、实验总结

本实验可以让学生了解 VRRP 的功能，熟悉 VRRP 的基本原理和配置，熟悉 VRRP 的工作过程。如果多个 VRRP 设备同时切换到主设备状态，则优先级较低的 VRRP 设备在 VRRP 通知消息交互协商后切换到备份状态，优先级最高的 VRRP 设备成为最终的主设备；如果优先级相同，则主 IP 地址数值较大的 VRRP 备份组将成为主设备。

六、实验思考题

（1）VRRP 的抢占模式和非抢占模式在什么情况下出现？
（2）VRRP 有几种状态机？分别是什么？
（3）backup 设备在该定时器超时后，仍未收到通告报文会变成什么状态？

OSPF 协议的高级配置与验证

实验二十

本书实验十七的主要内容是认识 OSPF 协议以及它的简单应用,实验内容涉及 OSPF 的单区域。在单区域中,每台路由器都要收集其他所有路由器的链路状态信息,随着网络规模的不断扩大,信息量也会不断增加,这样下去每台路由器的工作量会越来越大,从而导致性能下降,影响数据转发,不便于管理。

为了解决上述问题,OSPF 协议可以将整个自治系统划分为不同的区域。就像一个国家的国土面积过大,单方面管理很不方便,就会划分不同的省份来管理。链路状态信息只在区域内部泛洪,区域之间传递的只是路由条目而非链路状态信息,因此大大减小了路由器的负担,有效地把拓扑变化控制在区域内,由此提高了网络的稳定性。

OSPF 多区域的优点:提高了网络的扩展性,有利于组建更大规模的网络;分区域后,各区域独立管理效率高、收敛速度快。

一、实验目的

(1) 熟悉 OSPF 的概念和用途。
(2) 掌握 OSPF 多区域的配置。
(3) 掌握 OSPF 路由虚链路以及重分发的配置方法。
(4) 掌握 OSPF 认证的配置方法。
(5) 掌握路由传递过程包的更新变化。

二、实验环境

eNSP V100R003C00SPC100 模拟器(路由器、交换机、直通线、计算机)。

三、实验内容及步骤

(1) 搭建实验拓扑。
(2) 地址规划。
(3) 确定实验思路。
(4) 配置接口 IP 信息。
(5) 配置 OSPF 多区域。
(6) 配置 OSPF 虚链路重分发等。
(7) 配置 OSPF 认证。

（8）配置命令。
（9）网络测试。

四、实验过程

1. 实验拓扑

结合实验内容，设计合理的网络拓扑，如图 20-1 所示。

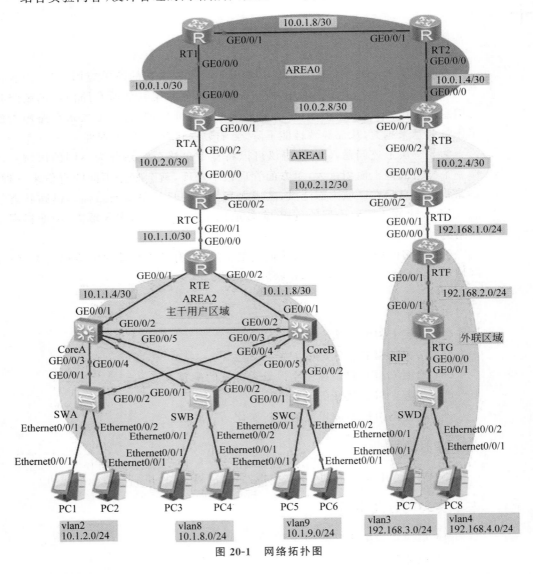

图 20-1　网络拓扑图

2. 地址规划

参照拓扑图，合理规划 IP 地址，如表 20-1 所示。

实验二十 OSPF 协议的高级配置与验证

表 20-1 IP 地址规划表

序号	设备	接口	所属 VLAN	IP 地址
1	PC1	FastEthernet0	2	10.1.2.0/24
2	PC2	FastEthernet0	2	10.1.2.0/24
3	PC3	FastEthernet0	8	10.1.8.0/24
4	PC4	FastEthernet0	8	10.1.8.0/24
5	PC5	FastEthernet0	9	10.1.9.0/24
6	PC6	FastEthernet0	9	10.1.9.0/24
7	PC7	FastEthernet0	3	192.168.3.0/24
8	PC8	FastEthernet0	4	192.168.4.0/24

3. 实验配置

（1）配置接口 IP 地址，如图 20-2 所示。

```
[RTA] sysname RTA
[RTA]interface GigabitEthernet0/0/0
[RTA-GigabitEthernet0/0/0] ip address 10.0.1.2 255.255.255.252
[RTA-GigabitEthernet0/0/0]quit
[RTA]interface GigabitEthernet0/0/1
[RTA-GigabitEthernet0/0/1] ip address 10.0.2.9 255.255.255.252
[RTA-GigabitEthernet0/0/1]quit
[RTA]interface GigabitEthernet0/0/2
[RTA-GigabitEthernet0/0/2] ip address 10.0.2.1 255.255.255.252
[RTA-GigabitEthernet0/0/2]quit
```

图 20-2 接口 IP 配置

（2）配置 OSPF 多区域，如图 20-3 所示。

```
[RTA]ospf 1
[RTA-ospf-1-area-0.0.0.0]  network 10.0.1.0 0.0.0.3
[RTA-ospf-1-area-0.0.0.0]  network 10.0.2.8 0.0.0.3
[RTA-ospf-1-area-0.0.0.0] area 0.0.0.1
[RTA-ospf-1-area-0.0.0.1]  network 10.0.2.0 0.0.0.3
```

图 20-3 OSPF 多区域配置

（3）配置 OSPF 虚链路重分发，如图 20-4 所示。

```
[RTA]ospf 1
[RTA-ospf-1-area-0.0.0.1]  vlink-peer 10.0.2.2 simple cipher 123
[RTA]ospf 1
[RTA-ospf-1-area-0.0.0.0]  authentication-mode simple cipher 123
```

图 20-4 OSPF 虚链路重分发配置

（4）区域 0（AREA0）路由器配置，如图 20-5 所示。

```
interface GigabitEthernet0/0/0 ip address 10.0.1.2 255.255.255.252    //配置接口 IP 地址
interface GigabitEthernet0/0/1 ip address 10.0.2.9 255.255.255.252
interface GigabitEthernet0/0/2 ip address 10.0.2.1 255.255.255.252
ospf 1                              //配置 OSPF
area 0.0.0.0
authentication-mode simple cipher %$%$R#YM8WYo=)*c<KT3WKsFfl6@%$%$    //配置
                                                                     //认证
network 10.0.1.0 0.0.0.3            //宣告广播地址
network 10.0.2.8 0.0.0.3
area 0.0.0.1
network 10.0.2.0 0.0.0.3
vlink-peer 10.0.2.2 simple cipher %$%$V&UH47is},fOe:En\ZdLf%_:%$%$    //配置虚链路
                                                                     //与认证
```

图 20-5　OSPF 区域 0 路由器配置

（5）区域 1（AREA1）路由器配置，如图 20-6 所示。

```
interface GigabitEthernet0/0/0 ip address 10.0.2.2 255.255.255.252     //配置接口 IP 地址
interface GigabitEthernet0/0/1 ip address 10.1.1.1 255.255.255.252
interface GigabitEthernet0/0/2 ip address 10.0.2.13 255.255.255.252
ospf 1                              //配置 OSPF 进程
area 0.0.0.1
network 10.0.2.0 0.0.0.3
network 10.0.2.12 0.0.0.3
vlink-peer 10.0.1.2 simple cipher %$%$z]!xRCo^\/0ZE1r6rE-9Tf$Uo%$%$    //配置虚链路
area 0.0.0.2
network 10.1.1.0 0.0.0.3
```

图 20-6　OSPF 区域 1 路由器配置

（6）外联区域路由器配置，如图 20-7 所示。

```
interface GigabitEthernet0/0/0 ip address 10.0.2.6 255.255.255.252    //配置接口 IP 地址
interface GigabitEthernet0/0/1 ip address 192.168.1.1 255.255.255.0
interface GigabitEthernet0/0/2 ip address 10.0.2.14 255.255.255.252
ospf 1                              //配置 OSPF
asbr-summary 192.168.0.0 255.255.0.0    //汇总路由
import-route rip 1                  //重分发 RIP
area 0.0.0.1
network 10.0.2.4 0.0.0.3
network 10.0.2.12 0.0.0.3
rip 1                               //配置 RIP
default-route originate             //下发默认路由
version 2                           //版本
network 192.168.1.0                 //宣告业务网段
import-route ospf 1                 //重分发 OSPF
```

图 20-7　OSPF 外联区域路由器配置

（7）核心交换机配置。

① 创建 VLAN，开启 DHCP 并配置地址池，如图 20-8 所示。

② 配置 SVI 接口、VRRP，配置 VRRP 虚拟地址，设置优先级，如图 20-9 所示；配置接

口类型;配置 OSPF,如图 20-10 所示。

```
interface Vlanif1
    ip address 10.1.1.6 255.255.255.252
interface Vlanif2
    ip address 10.1.2.2 255.255.255.0
    vrrp vrid 2 virtual-ip 10.1.2.1
    vrrp vrid 2 priority 120
interface Vlanif8
    ip address 10.1.8.2 255.255.255.0
    vrrp vrid 8 virtual-ip 10.1.8.1
    vrrp vrid 8 priority 120
    dhcp select global
interface Vlanif9
    ip address 10.1.9.2 255.255.255.0
    vrrp vrid 9 virtual-ip 10.1.9.1
    vrrp vrid 9 priority 120
    dhcp select global
interface GigabitEthernet0/0/1
    port link-type access
interface GigabitEthernet0/0/2
    port link-type trunk
    port trunk allow-pass vlan 2 to 4094
interface GigabitEthernet0/0/3
    port link-type trunk
    port trunk allow-pass vlan 2 to 4094
interface GigabitEthernet0/0/4
    port link-type trunk
    port trunk allow-pass vlan 2 to 4094
interface GigabitEthernet0/0/5
    port link-type trunk
    port trunk allow-pass vlan 2 to 4094
```

```
vlan batch 2 to 10
dhcp enable
ip pool 8
    gateway-list 10.1.8.1
    network 10.1.8.0 mask 255.255.255.0
    dns-list 10.1.2.101
ip pool 9
    gateway-list 10.1.9.1
    network 10.1.9.0 mask 255.255.255.0
    dns-list 10.1.2.101
```

图 20-8 地址池配置　　　　　图 20-9 核心交换机配置 VRRP

```
ospf 1
    area 0.0.0.2
    network 10.1.1.4 0.0.0.3
    network 10.1.2.0 0.0.0.255
    network 10.1.8.0 0.0.0.255
    network 10.1.9.0 0.0.0.255
```

图 20-10 核心交换机配置 OSPF

③ 接入交换机配置。创建 VLAN、配置接口 IP,如图 20-11 所示。

```
vlan batch 2 to 10
interface Ethernet0/0/1
    port link-type access
    port default vlan 2
    stp edged-port enable
interface Ethernet0/0/2
    port link-type access
    port default vlan 2
    stp edged-port enable
interface Ethernet0/0/3
    port link-type access
    port default vlan 2
    stp edged-port enable
interface GigabitEthernet0/0/1
    port link-type trunk
    port trunk allow-pass vlan 2 to 4094
interface GigabitEthernet0/0/2
    port link-type trunk
    port trunk allow-pass vlan 2 to 4094
```

图 20-11 接入交换机配置

4. 结果验证

（1）测试用户自动获取 IP 地址，如图 20-12 所示。

图 20-12 测试 PC5 自动获取 IP 地址

（2）测试用户互访，如图 20-13 所示。

```
PC5
基础配置  命令行  组播  UDP发包工具  串口
Link local IPv6 address...........: fe80::5689:98ff:feee:3a18
IPv6 address......................: :: / 128
IPv6 gateway......................: ::
IPv4 address......................: 10.1.8.253
Subnet mask.......................: 255.255.255.0
Gateway...........................: 10.1.8.1
Physical address..................: 54-89-98-EE-3A-18
DNS server........................: 10.1.2.101


PC>ping 10.1.9.254

Ping 10.1.9.254: 32 data bytes, Press Ctrl_C to break
From 10.1.9.254: bytes=32 seq=1 ttl=127 time=125 ms
From 10.1.9.254: bytes=32 seq=2 ttl=127 time=125 ms
From 10.1.9.254: bytes=32 seq=3 ttl=127 time=125 ms
From 10.1.9.254: bytes=32 seq=4 ttl=127 time=110 ms
From 10.1.9.254: bytes=32 seq=5 ttl=127 time=140 ms

--- 10.1.9.254 ping statistics ---
  5 packet(s) transmitted
  5 packet(s) received
  0.00% packet loss
  round-trip min/avg/max = 110/125/140 ms
```

图 20-13　测试用户互访

(3) 查看 VRRP 状态，如图 20-14 所示。

```
<CoreA>display vrrp brief
VRID  State        Interface                   Type     Virtual IP
----------------------------------------------------------------
2     Master       Vlanif2                     Normal   10.1.2.1
8     Master       Vlanif8                     Normal   10.1.8.1
9     Master       Vlanif9                     Normal   10.1.9.1
----------------------------------------------------------------
Total:3     Master:3     Backup:0     Non-active:0
```

图 20-14　查看 VRRP 状态

(4) 查看生成树状态，如图 20-15 所示。

```
<CoreA>display stp brief
 MSTID  Port                        Role  STP State     Protection
   0    GigabitEthernet0/0/1        DESI  FORWARDING      NONE
   0    GigabitEthernet0/0/2        ALTE  DISCARDING      NONE
   0    GigabitEthernet0/0/3        DESI  FORWARDING      NONE
   0    GigabitEthernet0/0/4        DESI  FORWARDING      NONE
   0    GigabitEthernet0/0/5        ROOT  FORWARDING      NONE
```

图 20-15　查看生成树状态

(5) 查看 OSPF 邻居，如图 20-16 所示。

```
<RTA>display ospf peer br
<RTA>display ospf peer brief

	 OSPF Process 1 with Router ID 10.0.1.2
		  Peer Statistic Information
 ----------------------------------------------------------------------------
 Area Id          Interface                        Neighbor id      State
 0.0.0.0          GigabitEthernet0/0/0             10.0.1.1         Full
 0.0.0.0          GigabitEthernet0/0/1             10.0.1.6         Full
 0.0.0.1          GigabitEthernet0/0/2             10.0.2.2         Full
 ----------------------------------------------------------------------------
```

图 20-16　查看 OSPF 邻居

(6) 查看 OSPF 状态，如图 20-17 所示。

```
<RTA>display ospf brief
        OSPF Process 1 with Router ID 10.0.1.2
              OSPF Protocol Information

RouterID: 10.0.1.2          Border Router:  AREA
Multi-VPN-Instance is not enabled
Global DS-TE Mode: Non-Standard IETF Mode
Graceful-restart capability: disabled
Helper support capability : not configured
Applications Supported: MPLS Traffic-Engineering
Spf-schedule-interval: max 10000ms, start 500ms, hold 1000ms
Default ASE parameters: Metric: 1 Tag: 1 Type: 2
Route Preference: 10
ASE Route Preference: 150
SPF Computation Count: 34
RFC 1583 Compatible
Retransmission limitation is disabled
Area Count: 2    Nssa Area Count: 0
ExChange/Loading Neighbors: 0
Process total up interface count: 4
Process valid up interface count: 3

Area: 0.0.0.0          (MPLS TE not enabled)
```

图 20-17　查看 OSPF 状态

（7）查看虚链路，如图 20-18 所示。

```
<RTA>display ospf vlink ?
  |     Matching output
  <cr>  Please press ENTER to execute command
<RTA>display ospf vlink

        OSPF Process 1 with Router ID 10.0.1.2
              Virtual Links

 Virtual-link Neighbor-id  -> 10.0.2.2, Neighbor-State: Full

 Interface: 10.0.2.1 (GigabitEthernet0/0/2)
 Cost: 1  State: P-2-P  Type: Virtual
 Transit Area: 0.0.0.1
 Timers: Hello 10 , Dead 40 , Retransmit 5 , Transmit Delay 1
 GR State: Normal
```

图 20-18　查看虚链路

（8）查看路由汇总，如图 20-19 所示。

```
             10.1.1.0/30   OSPF    10   3       D   10.0.2.9    GigabitEthernet
0/0/1
             10.1.1.4/30   OSPF    10   4       D   10.0.2.9    GigabitEthernet
0/0/1
             10.1.1.8/30   OSPF    10   6       D   10.0.2.9    GigabitEthernet
0/0/1
             10.1.2.0/24   OSPF    10   5       D   10.0.2.9    GigabitEthernet
0/0/1
             10.1.2.1/32   OSPF    10   5       D   10.0.2.9    GigabitEthernet
0/0/1
             10.1.8.0/24   OSPF    10   5       D   10.0.2.9    GigabitEthernet
0/0/1
             10.1.8.1/32   OSPF    10   5       D   10.0.2.9    GigabitEthernet
0/0/1
             10.1.9.0/24   OSPF    10   5       D   10.0.2.9    GigabitEthernet
0/0/1
             10.1.9.1/32   OSPF    10   5       D   10.0.2.9    GigabitEthernet
0/0/1
             192.168.0.0/16 O_ASE  150  2       D   10.0.2.6    GigabitEthernet
0/0/2

OSPF routing table status : <Inactive>
        Destinations : 0      Routes : 0
```

图 20-19　查看路由汇总

（9）查看 RIP 表，如图 20-20 所示。

```
<RTF>display ip routing-table protocol rip
Route Flags: R - relay, D - download to fib
------------------------------------------------------------------------------
Public routing table : RIP
         Destinations : 19       Routes : 19

RIP routing table status : <Active>
         Destinations : 19       Routes : 19

Destination/Mask    Proto   Pre  Cost      Flags NextHop         Interface

        0.0.0.0/0   RIP     100  1           D   192.168.1.1     GigabitEthern
0/0/0
       10.0.1.0/30  RIP     100  1           D   192.168.1.1     GigabitEthern
0/0/0
       10.0.1.4/30  RIP     100  1           D   192.168.1.1     GigabitEthern
0/0/0
       10.0.1.8/30  RIP     100  1           D   192.168.1.1     GigabitEthern
0/0/0
       10.0.2.0/30  RIP     100  1           D   192.168.1.1     GigabitEthern
0/0/0
       10.0.2.4/30  RIP     100  1           D   192.168.1.1     GigabitEthern
0/0/0
       10.0.2.8/30  RIP     100  1           D   192.168.1.1     GigabitEthern
0/0/0
      10.0.2.12/30  RIP     100  1           D   192.168.1.1     GigabitEthern
0/0/0
       10.1.1.0/30  RIP     100  1           D   192.168.1.1     GigabitEthern
0/0/0
       10.1.1.4/30  RIP     100  1           D   192.168.1.1     GigabitEthern
0/0/0
       10.1.1.8/30  RIP     100  1           D   192.168.1.1     GigabitEthern
```

图 20-20　查看 RIP 表

（10）查看 OSPF Hello 报文，如图 20-21 所示。

```
应用显示过滤器 … <Ctrl-/>
No.  Time        Source     Destination  Protocol  Length  Info
  1  0.000000    10.0.1.2   224.0.0.5    OSPF      82      Hello Packet
  2  2.546000    10.0.1.1   224.0.0.5    OSPF      82      Hello Packet
  3  9.265000    10.0.1.2   224.0.0.5    OSPF      82      Hello Packet
  4  11.703000   10.0.1.1   224.0.0.5    OSPF      82      Hello Packet
  5  18.515000   10.0.1.2   224.0.0.5    OSPF      82      Hello Packet
  6  20.875000   10.0.1.1   224.0.0.5    OSPF      82      Hello Packet
  7  27.781000   10.0.1.2   224.0.0.5    OSPF      82      Hello Packet
  8  30.031000   10.0.1.1   224.0.0.5    OSPF      82      Hello Packet

> Frame 1: 82 bytes on wire (656 bits), 82 bytes captured (656 bits) on interface -, id 0
> Ethernet II, Src: HuaweiTe_05:08:5f (00:e0:fc:05:08:5f), Dst: IPv4mcast_05 (01:00:5e:00:00:05)
> Internet Protocol Version 4, Src: 10.0.1.2, Dst: 224.0.0.5
v Open Shortest Path First
  v OSPF Header
      Version: 2
      Message Type: Hello Packet (1)
      Packet Length: 48
      Source OSPF Router: 10.0.1.2
      Area ID: 0.0.0.0 (Backbone)
      Checksum: 0xcf97 [correct]
      Auth Type: Simple password (1)
      Auth Data (Simple): 123
  v OSPF Hello Packet
      Network Mask: 255.255.255.252
      Hello Interval [sec]: 10
    > Options: 0x02, (E) External Routing
      Router Priority: 1
      Router Dead Interval [sec]: 40
      Designated Router: 10.0.1.2
      Backup Designated Router: 10.0.1.1
      Active Neighbor: 10.0.1.1
```

图 20-21　查看 OSPF Hello 报文

Hello 报文用来建立、保持邻居关系和 DR 的选举。由广播网络、点对点网络上所有路由器的所有接口周期性地发出。非广播网络上则从 DR 的所有接口周期性发出。如图 20-21 所示,设备 RTA 向设备 RT1 定时发出 Hello 报文保活。

下面对 Hello 报文进行解析。

- Network Mask(网络掩码,4 字节):发送 Hello 报文接口的子网掩码。
- Hello Interval(Hello 间隔,2 字节):发送 Hello 报文的时间间隔(默认 10s),虚链路上默认间隔为 30s。
- Options(选项,8 位):可选项(1 为允许,0 为不允许)。

① DN:用于避免 MPLS VPN 中 BGP 和 OSPF 重复通告而产生的环路,上述情境中,当 Type3、5、7 的 LSA 置为 1 后,接收路由器就不能将该条 LSA 用于计算 SPF。

② O:表示始发路由器是否支持 Type9~Type11 Opaque LSA。

③ DC:表示始发路由器是否为虚链路。

④ L:本地链路信息(link-local signaling,LLS)是否开启平滑重启(graceful restart, GR)技术。

⑤ NP:N 置位仅用于 Hello 报文,表示是否有能力接收和发送 Type7 LSA(NSSA External LSA),如果 N 置位为 1 则 E 置位必须为 0;P 置位仅用于 NSSA 的 LSA 中,表示 Type7 LSA 是否转换为 Type5 LSA。

⑥ MC:表示是否为 IP 组播 OSPF。

⑦ E:表示是否有能力接收和发送 Type5 LSA(AS-External-LSA,区域外部 LSA),该值在所有末节区域中为 0,在骨干区域、非末节区域和所有的外部 LSA 中为 1,如果设置错误,将无法形成邻接关系。

⑧ MT:表示始发路由器是否支持多拓扑 OSPF(MT-OSPF),多拓扑 OSPF 暂未广泛应用。

- Router Priority(路由器优先级,1 字节):用来发送 Hello 报文接口,可以用于选举 DR 和 BDR。
- Router Dead Interval(路由失效时间,4 字节):如果在这个时间内,还没有收到邻居发送来的 Hello 报文,那么就会认为邻居失效(默认为 40s,即 4 倍 Hello Interval)虚链路上默认间隔为 120s。
- Designated Router(DR 的 IP 地址,4 字节):指定 DR 的接口 IP 地址,未选举时为 0.0.0.0。
- Backup Designated Router(BDR 的 IP 地址,4 字节):用来指定 BDR 的接口 IP 地址,未选举时为 0.0.0.0。
- Active Neighbor(邻居,4 字节):包含该路由器所有邻居路由器的 Router ID 列表。

(11) 查看 OSPF LSU 报文,如图 20-22 所示。

LSU 报文用于响应邻居路由器发的 LSR。根据 LSR 请求列表,将相应的 LSA 发送给邻居路由器,实现 LSA 的泛洪和同步。如图 20-22 所示,设备 RTA 向设备 RT1 发出 LSU 报文更新路由。

- Number of LSAs(LSA 数量,4 字节):标识该数据包中所有 LSA 的数量信息。
- LSA(链路状态通告):LSA 更新信息(包含每一条具体的 LSA 信息)。

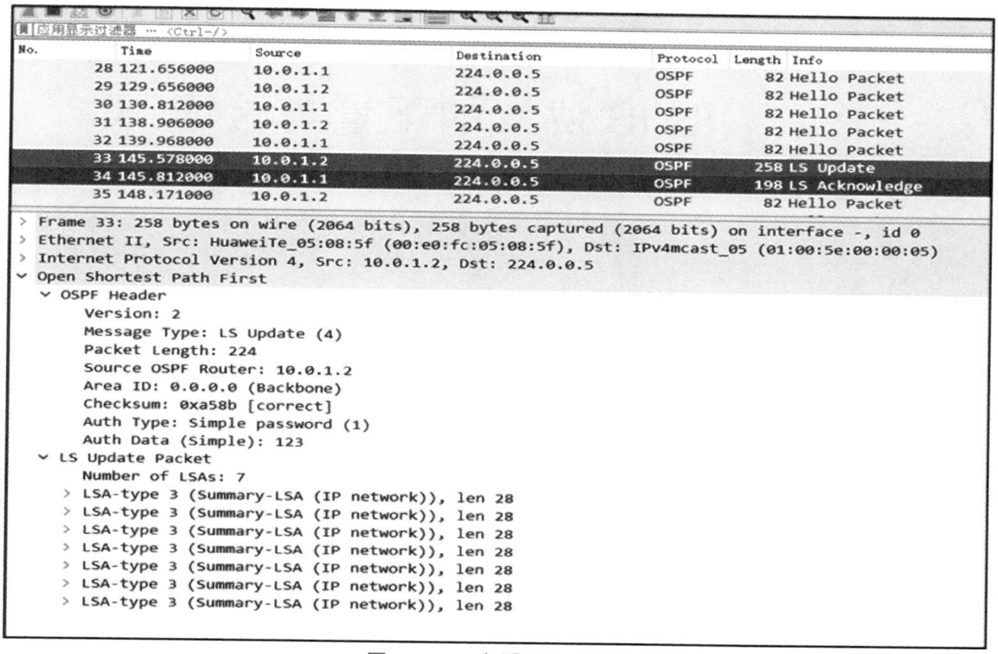

图 20-22　查看 LSU 报文

五、实验总结

本实验能够让学生进一步认识到 OSPF 技术在网络中的作用，认识到包传递过程中发生的变化，学会解析报文，学会配置多区域 OSPF、虚链路以及认证等功能，实现 OSPF 内部通信。

六、实验思考题

（1）OSPF 泛洪机制是什么？

（2）OSPF 老化机制是什么？

（3）ACL 分为哪几种？

实验二十一 IS-IS 协议的配置与验证

由于 OSPF 协议复杂、各种区域多、有限状态机多且拥有多种 LSA，因此出现了 IS-IS（intermediate system-to-intermediate system intra-domain routing information exchange protocol，中间系统到中间系统域内路由信息）。在采用 SPF 算法的基础上，ISO 对 OSPF 做出了很多简化，专门为 ISP 网络设计了链路状态型动态路由选择协议。IS-IS 即内部网关协议（interior gateway protocol，IGP），在自治系统内部使用。它是一种链路状态协议，使用最短路径优先（SPF）算法计算路由。

IS-IS 协议优先级为 15（华为）或 115（CISCO）。SPF 算法的每段链路开销均为 10。IS-IS 两个区域层次（有两个级别）：普通区域（Areas）叫 Level-1（L1）、骨干区（Backbone）叫 Level-2（L2）。

Level-1 的普通区域中所有的路由器必须有相同的区域地址，相互间形成 Level-1 的邻居关系，整个区域中只有 Level-1 层次上的链路数据库 LSDB，没有 Level-2 骨干区域的路由信息。

所有连续的 Level-2 路由器集合的区域叫骨干区，由所有的 L2（含 L1/L2）路由器组成，要求必须是可达的。

具体规则如下。

(1) Level-1 只能与本区域内 Level-1 或者 Level1-2 建邻。

(2) Level1-2 可以与任意区域相邻 Level-2 或者 L1-2 建邻，可以与同区域内 Level-1 建邻。

(3) Level-2 可以与任意区域相邻的 Level-2 或 Level1-2 建邻。

一、实验目的

(1) 理解 IS-IS 协议的基本原理。
(2) 掌握 IS-IS 的常用配置方法。
(3) 熟悉路由器的路由表。
(4) 熟悉 IS-IS 路由协议基本术语。

二、实验环境

eNSP V100R003C00SPC100 模拟器（路由器、交换机、直通线、计算机）。

三、实验内容及步骤

（1）搭建实验拓扑。
（2）配置接口 IP 信息。
（3）配置 IS-IS 协议。
（4）网络测试。
（5）查看 IS-IS 路由信息。

四、实验过程

1. 实验拓扑

根据实验内容，设计网络拓扑，如图 21-1 所示。

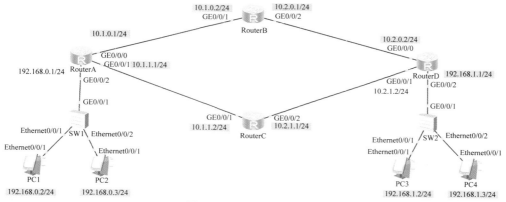

图 21-1　IS-IS 实验拓扑图

2. 地址规划

参照拓扑图，合理规划 IP 地址，如表 21-1 所示。

表 21-1　IP 地址规划表

设　　备	接　口	IP 地　址
RouterA	GE0/0/1	10.1.1.1/24
RouterA	GE0/0/2	192.168.0.1/24
RouterB	GE0/0/1	10.1.0.2/24
RouterB	GE0/0/2	10.2.0.1/24
RouterC	GE0/0/1	10.1.1.2/24
RouterC	GE0/0/2	10.2.1.1/24
RouterD	GE0/0/0	10.2.0.2/24
RouterD	GE0/0/1	10.2.1.2/24
RouterD	GE0/0/2	192.168.1.1/24
PC1	Eth0/0/1	192.168.0.2/24

续表

设　　备	接　　口	IP 地 址
PC2	Eth0/0/1	192.168.0.3/24
PC3	Eth0/0/1	192.168.1.2/24
PC4	Eth0/0/1	192.168.1.3/24

3. 实验配置

（1）对于 RouterA，配置 IS-IS 协议、配置 IS-IS 进程网络实体名称、配置接口 IP 地址、在接口上启用 IS-IS 协议并保存配置，如图 21-2 所示。

```
[RouterA]isis 1
[RouterA-isis-1]network-entity 10.0000.0000.0001.00
[RouterA]interface GigabitEthernet0/0/0
[RouterA-GigabitEthernet0/0/0]ip address 10.1.0.1 255.255.255.0
[RouterA-GigabitEthernet0/0/0]isis enable 1
[RouterA]interface GigabitEthernet0/0/1
[RouterA-GigabitEthernet0/0/1]ip address 10.1.1.1 255.255.255.0
[RouterA-GigabitEthernet0/0/1]isis enable 1
[RouterA]interface GigabitEthernet0/0/2
[RouterA-GigabitEthernet0/0/2]ip address 192.168.0.1 255.255.255.0
[RouterA-GigabitEthernet0/0/2]isis enable 1
<RouterA>save
```

图 21-2　配置 RouterA

（2）对于 RouterB，配置 IS-IS 协议、配置 IS-IS 进程网络实体名称、配置接口 IP 地址、在接口上启用 IS-IS 协议并保存配置，如图 21-3 所示。

```
[RouterB]isis 1
[RouterB-isis-1]network-entity 10.0000.0000.0002.00
[RouterB]interface GigabitEthernet0/0/1
[RouterB-GigabitEthernet0/0/1]ip address 10.1.0.2 255.255.255.0
[RouterB-GigabitEthernet0/0/1]isis enable 1
[RouterB]interface GigabitEthernet0/0/2
[RouterB-GigabitEthernet0/0/2]ip address 10.2.0.1 255.255.255.0
[RouterB-GigabitEthernet0/0/2]isis enable 1
<RouterB>save
```

图 21-3　配置 RouterB

（3）对于 RouterC，配置 IS-IS 协议及 IS-IS 进程网络实体名称、配置接口 IP 地址、在接口上启用 IS-IS 协议并保存配置，如图 21-4 所示。

（4）对于 RouterD，配置 IS-IS 协议及 IS-IS 进程网络实体名称、配置接口 IP 地址、在接口上启用 IS-IS 协议并保存配置，如图 21-5 所示。

4. 结果验证

（1）验证 RouterA 的 IS-IS 结果，如图 21-6 所示。

实验二十一 IS-IS 协议的配置与验证

```
[RouterC]isis 1
[RouterC-isis-1]network-entity 10.0000.0000.0003.00
[RouterC]interface GigabitEthernet0/0/1
[RouterC-GigabitEthernet0/0/1]ip address 10.1.1.2 255.255.255.0
[RouterC-GigabitEthernet0/0/1]isis enable 1
[RouterC]interface GigabitEthernet0/0/2
[RouterC-GigabitEthernet0/0/2]ip address 10.2.1.1 255.255.255.0
6[RouterC-GigabitEthernet0/0/2]isis enable 1
<RouterC>save
```

图 21-4 配置 RouterC

```
[RouterD]isis 1
[RouterD-isis-1]network-entity 20.0000.0000.0004.00
[RouterD]interface GigabitEthernet0/0/0
[RouterD-GigabitEthernet0/0/0]ip address 10.2.0.2 255.255.255.0
[RouterD-GigabitEthernet0/0/0]isis enable 1
[RouterD]interface GigabitEthernet0/0/1
[RouterD-GigabitEthernet0/0/1]ip address 10.2.1.2 255.255.255.0
[RouterD-GigabitEthernet0/0/1]isis enable 1
[RouterD]interface GigabitEthernet0/0/2
[RouterD-GigabitEthernet0/0/2]ip address 192.168.1.1 255.255.255.0
[RouterD-GigabitEthernet0/0/2]isis enable 1
<RouterD>save
```

图 21-5 配置 RouterD

```
RouterA
<RouterA>disp isis lsdb

                   Database information for ISIS(1)
                   -------------------------------

                        Level-1 Link State Database

LSPID                  Seq Num        Checksum     Holdtime     Length    ATT/P/OL
--------------------------------------------------------------------------------
0000.0000.0001.00-00*  0x0000000c     0x264b       1063         111       1/0/0
0000.0000.0002.00-00   0x00000009     0x1d5f       1151         84        1/0/0
0000.0000.0002.01-00   0x00000004     0x9ee8       1151         55        0/0/0
0000.0000.0003.00-00   0x00000008     0xc4b2       1074         84        1/0/0
0000.0000.0003.01-00   0x00000003     0xa9dc       1074         55        0/0/0

Total LSP(s): 5
    *(In TLV)-Leaking Route, *(By LSPID)-Self LSP, +-Self LSP(Extended),
     ATT-Attached, P-Partition, OL-Overload

                        Level-2 Link State Database

LSPID                  Seq Num        Checksum     Holdtime     Length    ATT/P/OL
--------------------------------------------------------------------------------
0000.0000.0001.00-00*  0x0000000a     0x75a9       1063         135       0/0/0
0000.0000.0002.00-00   0x00000009     0x86f1       1151         131       0/0/0
0000.0000.0002.01-00   0x00000002     0xa2e6       1151         55        0/0/0
0000.0000.0002.02-00   0x00000002     0xef95       1151         55        0/0/0
0000.0000.0003.00-00   0x0000000c     0xcda2       1074         131       0/0/0
0000.0000.0003.01-00   0x00000002     0xabdb       1074         55        0/0/0
0000.0000.0003.02-00   0x00000002     0xf88a       1074         55        0/0/0
0000.0000.0004.00-00   0x00000008     0x6ff0       1188         111       0/0/0

Total LSP(s): 8
    *(In TLV)-Leaking Route, *(By LSPID)-Self LSP, +-Self LSP(Extended),
     ATT-Attached, P-Partition, OL-Overload
```

图 21-6 RouterA 的 IS-IS 结果

（2）验证 RouterB 的 IS-IS 结果，如图 21-7 所示。

```
<RouterB>disp isis lsdb

            Database information for ISIS(1)
            --------------------------------

                Level-1 Link State Database

LSPID                 Seq Num       Checksum    Holdtime      Length   ATT/P/OL
---------------------------------------------------------------------------------
0000.0000.0001.00-00  0x0000000c    0x264b      935           111      1/0/0
0000.0000.0002.00-00* 0x00000009    0x1d5f      1024          84       1/0/0
0000.0000.0002.01-00* 0x00000004    0x9ee8      1024          55       0/0/0
0000.0000.0003.00-00  0x00000008    0xc4b2      945           84       1/0/0
0000.0000.0003.01-00  0x00000003    0xa9dc      945           55       0/0/0

Total LSP(s): 5
        *(In TLV)-Leaking Route, *(By LSPID)-Self LSP, +-Self LSP(Extended),
                ATT-Attached, P-Partition, OL-Overload


                Level-2 Link State Database

LSPID                 Seq Num       Checksum    Holdtime      Length   ATT/P/OL
---------------------------------------------------------------------------------
0000.0000.0001.00-00  0x0000000a    0x75a9      935           135      0/0/0
0000.0000.0002.00-00* 0x00000009    0x86f1      1024          131      0/0/0
0000.0000.0002.01-00* 0x00000002    0xa2e6      1024          55       0/0/0
0000.0000.0002.02-00* 0x00000002    0xef95      1024          55       0/0/0
0000.0000.0003.00-00  0x0000000c    0xcda2      945           131      0/0/0
0000.0000.0003.01-00  0x00000002    0xabdb      945           55       0/0/0
0000.0000.0003.02-00  0x00000002    0xf88a      945           55       0/0/0
0000.0000.0004.00-00  0x00000008    0x6ff0      1061          111      0/0/0

Total LSP(s): 8
        *(In TLV)-Leaking Route, *(By LSPID)-Self LSP, +-Self LSP(Extended),
                ATT-Attached, P-Partition, OL-Overload
```

图 21-7　RouterB 的 IS-IS 结果

（3）验证 RouterC 的 IS-IS 结果，如图 21-8 所示。

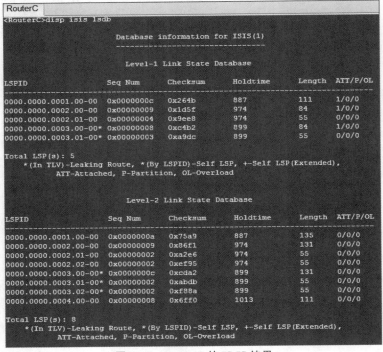

```
<RouterC>disp isis lsdb

            Database information for ISIS(1)
            --------------------------------

                Level-1 Link State Database

LSPID                 Seq Num       Checksum    Holdtime      Length   ATT/P/OL
---------------------------------------------------------------------------------
0000.0000.0001.00-00  0x0000000c    0x264b      887           111      1/0/0
0000.0000.0002.00-00  0x00000009    0x1d5f      974           84       1/0/0
0000.0000.0002.01-00  0x00000004    0x9ee8      974           55       0/0/0
0000.0000.0003.00-00* 0x00000008    0xc4b2      899           84       1/0/0
0000.0000.0003.01-00* 0x00000003    0xa9dc      899           55       0/0/0

Total LSP(s): 5
        *(In TLV)-Leaking Route, *(By LSPID)-Self LSP, +-Self LSP(Extended),
                ATT-Attached, P-Partition, OL-Overload


                Level-2 Link State Database

LSPID                 Seq Num       Checksum    Holdtime      Length   ATT/P/OL
---------------------------------------------------------------------------------
0000.0000.0001.00-00  0x0000000a    0x75a9      887           135      0/0/0
0000.0000.0002.00-00  0x00000009    0x86f1      974           131      0/0/0
0000.0000.0002.01-00  0x00000002    0xa2e6      974           55       0/0/0
0000.0000.0002.02-00  0x00000002    0xef95      974           55       0/0/0
0000.0000.0003.00-00* 0x0000000c    0xcda2      899           131      0/0/0
0000.0000.0003.01-00* 0x00000002    0xabdb      899           55       0/0/0
0000.0000.0003.02-00* 0x00000002    0xf88a      899           55       0/0/0
0000.0000.0004.00-00  0x00000008    0x6ff0      1013          111      0/0/0

Total LSP(s): 8
        *(In TLV)-Leaking Route, *(By LSPID)-Self LSP, +-Self LSP(Extended),
                ATT-Attached, P-Partition, OL-Overload
```

图 21-8　RouterC 的 IS-IS 结果

（4）验证 RouterD 的 IS-IS 结果，如图 21-9 所示。

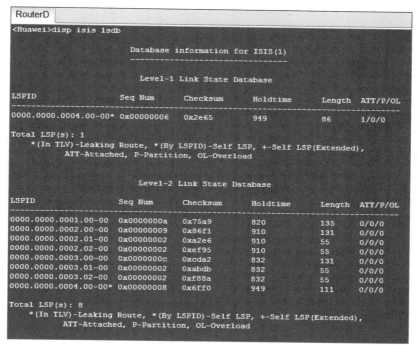

图 21-9　RouterD 的 IS-IS 结果

（5）测试从 PC1 到 PC3 和 PC4 是否正常。PC1 测试如图 21-10 所示。

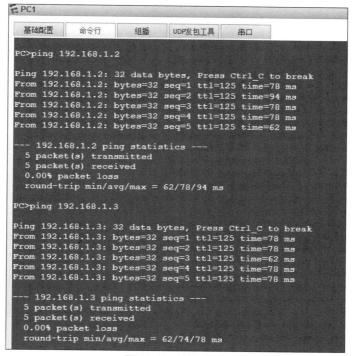

图 21-10　PC1 测试

五、实验总结

本实验可以让学生进一步了解 IS-IS 技术在网络中的应用,学会搭建 IS-IS 环境,实现通过 IS-IS 完成内部网络的通信。Level-1 只能与本区域内 Level-1 或者 Level1-2 建邻;Level1-2 可以与任意区域相邻 Level-2 或者 L1-2 建邻,可以与同区域内 Level-1 建邻;Level-2 可以与任意区域相邻的 Level-2 或 Level1-2 建邻。

六、实验思考题

(1) IS-IS 路由协议与 OSPF 协议的不同点在哪?
(2) IS-IS 有几种报文类型?
(3) IS-IS 是如何进行报文交互的?

实验二十二　交换机端口安全配置与验证

一、实验目的

(1) 理解交换机的 MAC 表。
(2) 理解交换机的端口安全特性。
(3) 学会交换机的端口安全配置。

二、实验环境

Cisco Packet Tracer 6.0 模拟器(路由器、交换机、直通线)。

三、实验内容及步骤

(1) 熟悉交换机端口安全的相关知识。
(2) 设计网络结构拓扑图。
(3) 对终端设备路由器进行配置。
(4) 进行交换机端口安全配置。
(5) 模拟非法接入。

四、实验过程

交换机的端口安全特性可以使当具有非法 MAC 地址的设备接入时,交换机会自动关闭接口或者拒绝非法设备接入,也可以限制某个端口上最大的 MAC 地址数,本实验限制 f0/1 接口只允许 R1 接入。

1. 实验拓扑

结合实验内容,主要考察交换机的端口安全,因此用 1 台路由器模拟终端设备的接入,具体拓扑如图 22-1 所示。

图 22-1　实验拓扑图

2. 路由器 R1 的配置与接口 MAC 地址查看

具体命令内容如下:

```
R1(config)#int g0/0
R1(config-if)#no shutdown
R1(config-if)#ip address 172.16.0.101 255.255.0.0
R1#show int g0/0
GigabitEthernet0/0 is up, line protocol is upHardware is MV96340 Ethernet,
address is 0019.5535.b828 (bia 0019.5535.b828) Internet address is 172.16.0.101/
16MTU 1500 bytes, BW 100000 Kbit, DLY 100 usec.
```

由上述内容可以看到 g0/0 接口的 MAC 地址是 0019.5535.b828。

3. 配置交换机端口安全

具体命令内容如下：

```
S1(config)#int f0/1
S1(config-if)#shutdown
S1(config-if)#switch mode access
//以上命令把端口改为访问模式，即用来接入计算机，在实验二十三将详细介绍该命令的含义
S1(config-if)#switch port-security
//以上命令是打开交换机的端口安全功能
S1(config-if)#switch port-security maximum 1
//以上命令只允许该端口下的 MAC 条目最大数量为 1，即只允许一台设备接入
S1(config-if)#switch port-security violation { protect | shutdown | restrict }
//protect 表示当新的计算机接入时，如果该接口的 MAC 条目超过最大数量，则这台新的计算机
//将无法接入，而原有的计算机不受影响；shutdown 表示当新的计算机接入时，如果该接口的 MAC
//条目超过最大数量，则该接口将会被关闭，导致这台新的计算机和原有的计算机都无法接入，需要
//管理员使用"no shutdown"命令重新打开；restrict 表示当新的计算机接入时，如果该接口的
//MAC 条目超过最大数量，则这台新的计算机可以接入，然后向交换机发送警告信息
S1(config-if)#switchport port-security mac-address 0019.5535.b828
//允许 R1 路由器从 f0/1 接口接入
S1(config-if)#no shutdown
S1(config)#int vlan1
S1(config-if)#no shutdown
S1(config-if)#ip address 172.16.0.1 255.255.0.0
//以上命令配置交换机的管理地址
```

4. 检查 MAC 地址表

具体命令内容如下：

```
S1#show mac-address-table
Mac Address Table
-------------------------------------------
Vlan    Mac Address       Type        Ports
----    -----------       --------    -----
All     0100.0ccc.cccc    STATIC      CPU
1       0018.ba11.eb91    DYNAMIC     Fa0/15
```

```
1   0019.5535.b828    STATIC    Fa0/1
Total Mac Addresses for this criterion: 24
//R1 的 MAC 地址已经被登记在 f0/1 接口，并且表明是静态加入的
```

5. 模拟非法接入

此时从 R1 ping 交换机的管理地址，可以 ping 通，具体命令内容如下：

```
R1#ping 172.16.0.1
Type escape sequence to abort.
Sending 5, 100-byte ICMP Echos to 172.16.0.1, timeout is 2 seconds:
!!!!!
Success rate is 100 percent (5/5), round-trip min/avg/max = 1/1/4 ms
```

在 R1 上修改 g0/0 的 MAC 地址为另一个地址，以此模拟另外一台设备的接入，具体命令内容如下：

```
R1(config)#int g0/0
R1(config-if)#mac-address 12.12.12
```

几秒后，则在 S1 上会出现以下内容：

```
01:09:39: %PM-4-ERR_DISABLE: psecure-violation error detected on Fa0/1, putting Fa0/1 in
err-disable state
01:09:39: %PORT_SECURITY-2-PSECURE_VIOLATION: Security violation occurred, caused by MAC address
0012.0012.0012 on port FastEthernet0/1.
01:09:40: %LINEPROTO-5-UPDOWN: Line protocol on Interface FastEthernet0/1, changed state to down
//以上内容提示 f0/1 接口被关闭
S1#show int f0/1
FastEthernet0/1 is down, line protocol is down (err-disabled)
Hardware is Fast Ethernet, address is 0018.ba11.f503 (bia 0018.ba11.f503)
MTU 1500 bytes, BW 100000 Kbit, DLY 100 usec,
reliability 255/255, txload 1/255, rxload 1/255
//以上内容表明 f0/1 接口因为错误而被关闭。非法设备移除后，在 f0/1 接口下，执行 shutdown
//和 no shutdown 命令可以重新打开该接口
```

6. 查看结果

具体命令内容如下：

```
S1#show port-security
Secure Port MaxSecureAddr CurrentAddr SecurityViolation Security Action
            (Count)       (Count)     (Count)
---------------------------------------------------------------------------
```

```
        Fa0/1 1 1 0 Shutdown
---------------------------------------------------------------
Total Addresses in System (excluding one mac per port) : 0
Max Addresses limit in System (excluding one mac per port) : 6272
//以上命令可以查看端口安全的设置情况
```

五、实验总结

配置交换机的端口安全实验，可以让学生了解到端口安全的重要性。端口安全能够保护内网不受到威胁，提升网络可靠性，禁止私接网络设备。

六、实验思考题

（1）如果上行交换机上设置的最大连接数为5，下行交换机连接了10台计算机，请描述一下网络的状态是怎样的。

（2）交换机的端口安全主要应用于生活中的哪些地方？

实验二十三 访问控制列表的配置与验证

访问控制列表(access control list,ACL)是路由器和交换机接口的指令列表,用来控制进出端口的数据包。ACL可以过滤网络中的流量,是控制访问的一种网络技术手段。配置ACL后,可以限制网络流量,允许特定设备访问,指定转发特定端口数据包等。例如,禁止局域网内的设备访问外部公共网络,或者只能使用某个服务。ACL既可以在路由器上配置,也可以在具有ACL功能的业务软件上进行配置。

ACL基本原理。ACL负责管理用户配置的所有规则,并提供报文匹配规则的算法。

ACL的规则匹配过程。当设备收到报文后,从报文中取出信息,形成搜索关键字值。将键值与ACL规则进行对比,只要有报文规则与之匹配,则停止搜索,将其称为命中规则;若查找完所有规则,条件无法匹配,则称为未命中规则。

ACL的规则分为"permit"(允许)规则和"deny"(拒绝)规则。

ACL可以将报文分成3类:命中permit规则的报文、命中deny规则的报文以及未命中规则的报文。

一、实验目的

(1) 掌握ACL设计原则和工作过程。
(2) 学会定义标准ACL和扩展ACL。
(3) 学会应用标准ACL和扩展ACL。
(4) 掌握标准ACL和扩展ACL的调试方法。

二、实验环境

Cisco Packet Tracer 6.0模拟器(路由器、计算机、直通线、串口线)。

三、实验内容及步骤

(1) 路由器基本配置。
(2) 路由配置。
(3) 访问控制列表配置。

四、实验过程

1. 构建实验拓扑

本实验共有两个基本任务,分别是标准 ACL 和扩展 ACL 的应用,为了更好地验证实验效果,在两个任务中采用相同的拓扑,如图 23-1 所示。

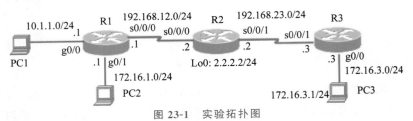

图 23-1 实验拓扑图

标准 ACL 的应用,主要目的是拒绝 PC2 所在网段的主机访问路由器 R2,同时只允许主机 PC3 访问路由器 R2,并在 PC3 上实现对路由器 R2 的远程管理,整个网络采用 EIGRP 保证网络的连通性。

扩展 ACL 的应用,主要目的是只允许 PC2 所在网段的主机访问路由器 R2 的 WWW 和 telnet 服务,并拒绝 PC3 所在网段的主机 ping 路由器 R2,整个网络同样采用 EIGRP 保证网络的连通性。

2. 标准 ACL 的配置与应用

1) 配置路由器 R1

具体命令内容如下:

```
R1(config)#router eigrp 1
R1(config-router)#network 10.1.1.0 0.0.0.255
R1(config-router)#network 172.16.1.0 0.0.0.255
R1(config-router)#network 192.168.12.0
R1(config-router)#no auto-summary
```

2) 配置路由器 R2

具体命令内容如下:

```
R2(config)#router eigrp 1
R2(config-router)#network 2.2.2.0 0.0.0.255
R2(config-router)#network 192.168.12.0
R2(config-router)#network 192.168.23.0
R2(config-router)#no auto-summary
R2(config)#access-list 1 deny 172.16.1.0 0.0.0.255      //定义 ACL
R2(config)#access-list 1 permit any
R2(config)#interface Serial0/0/0
R2(config-if)#ip access-group 1 in                       //在接口上应用 ACL
R2(config)#access-list 2 permit 172.16.3.1
R2(config-if)#line vty 0 4
```

```
R2(config-line)#access-class 2 in        //在 vty 下应用 ACL
R2(config-line)#password cisco
R2(config-line)#login
```

3）配置路由器 R3

具体命令内容如下：

```
R3(config)#router eigrp 1
R3(config-router)#network 172.16.3.0 0.0.0.255
R3(config-router)#network 192.168.23.0
R3(config-router)#no auto-summary
```

关于以上命令的技术说明如下。

（1）将 ACL 定义好，可以在很多地方应用，接口上的应用只是其中之一，其他的常用应用包括在 route-map 中的 match 应用以及在 vty 下用"access-class"命令调用，来控制 telnet 的访问。

（2）访问控制列表表项的检查按自上而下的顺序进行，并且从第一个表项开始，因此必须考虑在访问控制列表中定义语句的次序。

（3）路由器不对自身产生的 IP 数据包进行过滤。

（4）访问控制列表最后一条是隐含地拒绝所有访问流量。

（5）每一个路由器接口的每一个方向，每一种协议只能创建一个 ACL。

（6）"access-class"命令只对标准 ACL 有效。

3. 标准 ACL 实验调试

由于在路由器 R2 上拒绝网段 176.16.1.0 的数据，所以 PC2 不能访问路由器 R2 的回环测试地址 2.2.2.2，从而验证了 PC2 所在网段不能访问路由器 R2；同理，在路由器 R2 上允许了 PC3 的数据通过。所以，当 R2 开启远程管理功能后，就可以通过 PC3 远程 telnet 服务地址 2.2.2.2 来管理 R2。

1）show ip access-lists

该命令用来查看所定义的 IP 访问控制列表。

```
R2#show ip access-lists
Standard IP access list 1
10 deny 172.16.1.0, wildcard bits 0.0.0.255 (11 matches)
20 permit any (405 matches)
Standard IP access list 2
10 permit 172.16.3.1 (2 matches)
```

以上输出内容表明路由器 R2 上定义的标准访问控制列表为"1"和"2"，括号中的数字表示匹配条件的数据包的个数，可以用"clear access-list counters"命令将访问控制列表计数器清零。

2) show ip interface

该命令用来查看接口的配置以及状态信息。

```
R2#show ip interface s0/0/0
Serial0/0/0 is up, line protocol is up
Internet address is 192.168.12.2/24
Broadcast address is 255.255.255.255
Address determined by setup command
MTU is 1500 bytes
Helper address is not set
Directed broadcast forwarding is disabled
Multicast reserved groups joined: 224.0.0.10
Outgoing access list is not set
Inbound access list is 1
...
```

以上输出内容表明在接口 s0/0/0 的入方向应用了访问控制列表 1。

4. 扩展 ACL 的配置与应用

由于在设计实验的时候,采用了同一张拓扑图,设备的连接方式、IP 地址的规划以及路由协议的选择都是一样的,因此,在配置扩展 ACL 时,就省略了关于前期接口配置和路由协议的配置过程,下面直接从扩展 ACL 开始配置。

1) 配置路由器 R1

具体命令内容如下:

```
R1(config)#access-list 100 permit tcp 172.16.1.0 0.0.0.255 host 2.2.2.2 eq www
R1(config)#access-list 100 permit tcp 172.16.1.0 0.0.0.255 host 192.168.12.2 eq www
R1(config)#access-list 100 permit tcp 172.16.1.0 0.0.0.255 host 192.168.23.2 eq www
R1(config)#access-list 100 permit tcp 172.16.1.0 0.0.0.255 host 2.2.2.2 eq telnet
R1(config)#access-list 100 permit tcp 172.16.1.0 0.0.0.255 host 192.168.12.2 eq telnet
R1(config)#access-list 100 permit tcp 172.16.1.0 0.0.0.255 host 192.168.23.2 eq telnet
R1(config)#interface g0/0
R1(config-if)#ip access-group 100 in
```

2) 配置路由器 R2

具体命令内容如下:

```
R2(config)#no access-list 1          //删除 ACL
R2(config)#no access-list 2
R2(config)#ip http server            //将路由器配置成 Web 服务器
R2(config)#line vty 0 4
R2(config-line)#password cisco
R2(config-line)#login
```

3）配置路由器 R3
具体命令内容如下：

```
R3(config)#access-list 101 deny icmp 172.16.3.0 0.0.0.255 host 2.2.2.2 log
R3(config)#access-list 101 deny icmp 172.16.3.0 0.0.0.255 host 192.168.12.2 log
R3(config)#access-list 101 deny icmp 172.16.3.0 0.0.0.255 host 192.168.23.2 log
R3(config)#access-list 101 permit ip any any
R3(config)#interface g0/0
R3(config-if)#ip access-group 101 in
```

关于以上命令的技术说明如下。

（1）参数 log 会生成相应的日志信息，用来记录经过 ACL 入口的数据包的情况。

（2）考虑尽量将扩展的访问控制列表放在靠近过滤源的位置上，这样创建的过滤器就不会反过来影响其他接口上的数据流。另外，尽量使标准的访问控制列表靠近目的网段，由于标准访问控制列表只使用源地址，如果将其靠近源会阻止数据包流向其他端口。

5. 扩展 ACL 实验调试

（1）分别在 PC2 上访问路由器 R2 的 telnet 和 WWW 服务，然后查看访问控制列表 100，命令内容如下：

```
R1#show ip access-lists
Extended IP access list 100
10 permit tcp 172.16.1.0 0.0.0.255 host 2.2.2.2 eq www (8 matches)
20 permit tcp 172.16.1.0 0.0.0.255 host 192.168.12.2 eq www
30 permit tcp 172.16.1.0 0.0.0.255 host 192.168.23.2 eq www
40 permit tcp 172.16.1.0 0.0.0.255 host 2.2.2.2 eq telnet (20 matches)
50 permit tcp 172.16.1.0 0.0.0.255 host 12.12.12.2 eq telnet (4 matches)
60 permit tcp 172.16.1.0 0.0.0.255 host 23.23.23.2 eq telnet (4 matches)
```

（2）在 PC3 所在网段的主机上 ping 路由器 R2，路由器 R3 上会出现下面的日志信息：

```
*Feb 25 17:35:46.383: %SEC-6-IPACCESSLOGDP: list 101 denied icmp 172.16.3.1 ->
2.2.2.2(0/0), 1 packet
*Feb 25 17:41:08.959: %SEC-6-IPACCESSLOGDP: list 101 denied icmp 172.16.3.1 ->
2.2.2.2(0/0), 4 packets
*Feb 25 17:42:46.919: %SEC-6-IPACCESSLOGDP: list 101 denied icmp 172.16.3.1 ->
192.168.12.2 (0/0), 1 packet
*Feb 25 17:42:56.803: %SEC-6-IPACCESSLOGDP: list 101 denied icmp 172.16.3.1 ->
192.168.23.2 (0/0), 1 packet
```

以上输出内容说明访问控制列表 101，当其在有匹配数据包的时候，系统更新了日志。

（3）在路由器 R3 上查看访问控制列表 101，命令内容如下：

```
R3#show access-lists
Extended IP access list 101
```

```
10 deny icmp 172.16.3.0 0.0.0.255 host 2.2.2.2 log (5 matches)
20 deny icmp 172.16.3.0 0.0.0.255 host 192.168.12.2 log (5 matches)
30 deny icmp 172.16.3.0 0.0.0.255 host 192.168.23.2 log (5 matches)
40 permit ip any any (6 matches)
```

五、实验总结

　　本实验可以让学生对 ACL 有一个较为全面的理解，并对日常生活中遇到的这方面的问题有一定的处理能力，为以后更深入地学习 ACL 的相关知识奠定一定的基础。

六、实验思考题

　　(1) 标准 ACL 和扩展 ACL 有什么区别？
　　(2) 标准 ACL 和扩展 ACL 的规则分别是什么？

实验二十四 防火墙安全策略的设计与配置

防火墙(firewall)是当前最重要的网络防护设备之一,它是位于两个及以上网络之间的组件,可以实现网络内部的访问控制。防火墙的本义是指在古代建造和使用木制房屋时,为防止火灾的发生和蔓延,人们用坚固的石块在房屋周围筑起的屏障。实际上,防火墙的作用与"门"相似。此"门"相当于安全规则,只有符合规则的人或事才被允许通过,与规则不匹配的则不被允许通过。

一、实验目的

(1) 熟悉防火墙的概念和用途。
(2) 掌握防火墙的基础配置方法。
(3) 掌握防火墙安全策略的配置方法。
(4) 掌握防火墙攻击防护的配置方法。

二、实验环境

eNSP V100R003C00SPC100 模拟器(路由器、直通线、计算机)、AR3260E-S 路由器 3 台、USG6000 防火墙 2 台、S5730 交换机 2 台、S3700 交换机 4 台、计算机 8 台。

三、实验内容及步骤

(1) 搭建实验拓扑。
(2) 配置接口 IP 信息。
(3) 配置域间策略以及 NAT 策略。
(4) 配置防火墙 HRP。
(5) 配置防火墙攻击防护。
(6) 网络测试。
(7) 配置防火墙攻击防护。
(8) 其他设备配置命令。
(9) 网络测试。

四、实验过程

1. 实验拓扑

根据实验内容,构建实验拓扑,如图 24-1 所示。

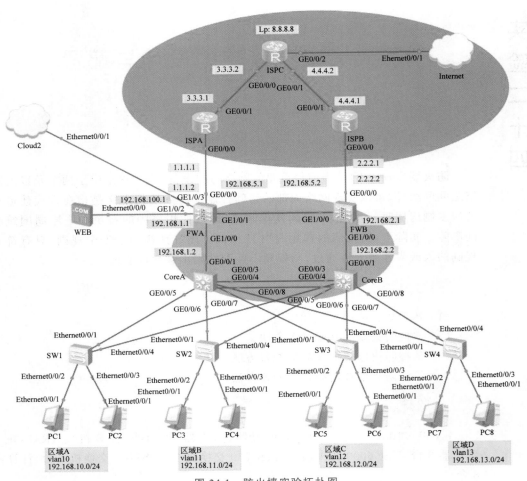

图 24-1 防火墙实验拓扑图

2. 地址规划

进行合理的 IP 地址规划，如表 24-1 所示。

表 24-1 IP 地址规划表

序 号	设 备	接 口	所属 VLAN	IP 地 址
1	PC1	FastEthernet0	10	192.168.10.0/24
2	PC2	FastEthernet0	10	192.168.10.0/24
3	PC3	FastEthernet0	11	192.168.11.0/24
4	PC4	FastEthernet0	11	192.168.11.0/24
5	PC5	FastEthernet0	12	192.168.12.0/24
6	PC6	FastEthernet0	12	192.168.12.0/24
7	PC7	FastEthernet0	13	192.168.13.0/24
8	PC8	FastEthernet0	13	192.168.13.0/24

3. 实验配置

（1）配置接口 IP 地址，如图 24-2 所示。

```
[FWA]sysname FWA
[FWA]interface GigabitEthernet0/0/0
[FWA-GigabitEthernet0/0/0] ip address 1.1.1.2 255.255.255.0
[FWA-GigabitEthernet0/0/0] alias GE0/METH
[FWA-GigabitEthernet0/0/0] service-manage ping permit
[FWA]interface GigabitEthernet1/0/0
[FWA-GigabitEthernet1/0/0] ip address 192.168.1.1 255.255.255.0
[FWA-GigabitEthernet1/0/0] service-manage https permit
[FWA-GigabitEthernet1/0/0] service-manage ping permit
[FWA-GigabitEthernet1/0/0] service-manage snmp permit
[FWA]interface GigabitEthernet1/0/1
[FWA-GigabitEthernet1/0/1] ip address 192.168.5.1 255.255.255.0
[FWA-GigabitEthernet1/0/1] service-manage ping permit
[FWA]interface GigabitEthernet1/0/2
[FWA-GigabitEthernet1/0/2] ip address 192.168.100.1 255.255.255.0
[FWA-GigabitEthernet1/0/2] service-manage ping permit
```

图 24-2　接口 IP 配置

（2）配置域间策略以及 NAT 策略，配置接口从属指定安全域，如图 24-3 所示。

```
[FWA]firewall zone trust
[FWA-zone-trust] set priority 85
[FWA-zone-trust] add interface GigabitEthernet1/0/0
[FWA-zone-trust] add interface GigabitEthernet1/0/1
[FWA]firewall zone untrust
[FWA-zone-untrust] set priority 5
[FWA-zone-untrust] add interface GigabitEthernet0/0/0
[FWA]firewall zone dmz
[FWA-zone-dmz] set priority 50
[FWA-zone-dmz] add interface GigabitEthernet1/0/2
[FWA]security-policy
[FWA-policy-security] rule name trust
[FWA-policy-security-rule-trust]  source-zone trust
[FWA-policy-security-rule-trust]  destination-zone dmz
[FWA-policy-security-rule-trust]  destination-zone untrust
[FWA-policy-security-rule-trust]  action permit
[FWA-policy-security-rule-trust] rule name dmz
[FWA-policy-security-rule-dmz]  source-zone dmz
[FWA-policy-security-rule-dmz]  destination-zone trust
[FWA-policy-security-rule-dmz]  action permit
[FWA-policy-security-rule-dmz] rule name untrust
[FWA-policy-security-rule-untrust]  source-zone untrust
[FWA-policy-security-rule-untrust]  destination-zone local
[FWA-policy-security-rule-untrust]  action permit
[FWA-policy-security]nat-policy
[FWA-policy-nat] rule name any
[FWA-policy-nat-rule-any]  source-zone trust
[FWA-policy-nat-rule-any]  destination-zone untrust
[FWA-policy-nat-rule-any]  action source-nat easy-ip
```

图 24-3　域间路由策略配置

(3) 配置防火墙 HRP,如图 24-4 所示。

```
HRP_M[FWA] hrp enable
HRP_M[FWA] hrp interface GigabitEthernet1/0/1 remote 192.168.5.2
HRP_M[FWA] hrp standby config enable
HRP_M[FWA] hrp track interface GigabitEthernet1/0/0
```

图 24-4　防火墙 HRP 配置

(4) 配置防火墙攻击防护,如图 24-5 所示。

```
HRP_M[FWA] firewall defend port-scan enable (+B)
HRP_M[FWA] firewall defend ip-sweep enable (+B)
HRP_M[FWA] firewall defend teardrop enable (+B)
HRP_M[FWA] firewall defend time-stamp enable (+B)
HRP_M[FWA] firewall defend route-record enable (+B)
HRP_M[FWA] firewall defend source-route enable (+B)
HRP_M[FWA] firewall defend ip-fragment enable (+B)
HRP_M[FWA] firewall defend tcp-flag enable (+B)
HRP_M[FWA] firewall defend winnuke enable (+B)
HRP_M[FWA] firewall defend fraggle enable (+B)
HRP_M[FWA] firewall defend tracert enable (+B)
HRP_M[FWA] firewall defend icmp-unreachable enable (+B)
HRP_M[FWA] firewall defend icmp-redirect enable (+B)
HRP_M[FWA] firewall defend large-icmp enable (+B)
HRP_M[FWA] firewall defend ping-of-death enable (+B)
HRP_M[FWA] firewall defend smurf enable (+B)
HRP_M[FWA] firewall defend land enable (+B)
HRP_M[FWA] firewall defend ip-spoofing enable (+B)
HRP_M[FWA] firewall defend action discard (+B)
```

图 24-5　防火墙攻击防护配置

(5) 配置防火墙 Web 黑名单,如图 24-6 所示。

```
HRP_M[FWA] file-frame web-reputation enable
HRP_M[FWA] file-frame web-reputation add black-host www.baidu.com
```

图 24-6　防火墙 Web 黑名单配置

(6) 配置最大新建连接速率,如图 24-7 所示。

```
HRP_M[FWA]interface GigabitEthernet0/0/0
HRP_M[FWA-GigabitEthernet0/0/0]firewall defend ipcar destination session-rate-limit
enable
```

图 24-7　最大新建连接速率配置

(7) 核心交换机 VLAN 和 Switch1 接口地址配置,如图 24-8 所示。

(8) 核心交换机上 DHCP 配置,创建地址池关联至对应 VLAN,配置网关地址、DNS、宣告所属网段,如图 24-9 所示。

(9) 配置地址获取方式,如图 24-10 所示。

(10) 核心交换机上配置 VRRP,如图 24-11 所示。

```
vlan batch 2 to 100
interface Vlanif1
    ip address 192.168.1.2 255.255.255.0
interface Vlanif3
    ip address 192.168.3.2 255.255.255.0
interface Vlanif10
    ip address 192.168.10.2 255.255.255.0
interface Vlanif11
    ip address 192.168.11.2 255.255.255.0
interface Vlanif12
    ip address 192.168.12.2 255.255.255.0
interface Vlanif13
    ip address 192.168.13.2 255.255.255.0
interface Vlanif100
    ip address 192.168.100.2 255.255.255.0
```

图 24-8　VLAN 的创建

```
ip pool 10
    gateway-list 192.168.10.1
    network 192.168.10.0 mask 255.255.255.0
    dns-list 192.168.100.101
ip pool 11
    gateway-list 192.168.11.1
    network 192.168.11.0 mask 255.255.255.0
    dns-list 192.168.100.101
ip pool 12
    gateway-list 192.168.12.1
    network 192.168.12.0 mask 255.255.255.0
    dns-list 192.168.100.101
ip pool 13
    gateway-list 192.168.13.1
    network 192.168.13.0 mask 255.255.255.0
    dns-list 192.168.100.101
```

图 24-9　地址池创建

```
interface Vlanif10
    dhcp select global
interface Vlanif11
    dhcp select global
interface Vlanif12
    dhcp select global
interface Vlanif13
    dhcp select global
```

图 24-10　地址获取方式配置

```
interface Vlanif10
    vrrp vrid 10 virtual-ip 192.168.10.1
    vrrp vrid 10 priority 120
interface Vlanif11
    vrrp vrid 11 virtual-ip 192.168.11.1
    vrrp vrid 11 priority 120
interface Vlanif12
    vrrp vrid 12 virtual-ip 192.168.12.1
    vrrp vrid 12 priority 120
interface Vlanif13
    vrrp vrid 13 virtual-ip 192.168.13.1
    vrrp vrid 13 priority 120
interface Vlanif100
    vrrp vrid 100 virtual-ip 192.168.100.1
    vrrp vrid 100 priority 120
```

图 24-11　VRRP 配置

（11）核心交换机上配置接口类型，如图 24-12 所示。

（12）核心交换机上配置链路聚合，并将指定接口加入链路聚合组，配置生成树根桥，如图 24-13 所示。

（13）接入交换机，配置接口类型，如图 24-14 所示。

（14）运营商配置，如图 24-15 所示。

（15）宣告 RIP，如图 24-16 所示。

4. 结果验证

（1）同网段间互访，如图 24-17 所示。

（2）客户端访问服务器，如图 24-18 所示。

```
interface GigabitEthernet0/0/1
    port link-type access
    stp edged-port enable
interface GigabitEthernet0/0/2
    port link-type access
    port default vlan 3
    stp edged-port enable
interface GigabitEthernet0/0/5
    port link-type trunk
    port trunk allow-pass vlan 2 to 4094
interface GigabitEthernet0/0/6
    port link-type trunk
    port trunk allow-pass vlan 2 to 4094
interface GigabitEthernet0/0/7
    port link-type trunk
    port trunk allow-pass vlan 2 to 4094
interface GigabitEthernet0/0/8
    port link-type trunk
    port trunk allow-pass vlan 2 to 4094
interface GigabitEthernet0/0/9
    port link-type trunk
    port trunk allow-pass vlan 2 to 4094
```

图 24-12 接口类型配置

```
interface Eth-Trunk1
    port link-type trunk
    port trunk allow-pass vlan 2 to 4094
interface GigabitEthernet0/0/3
    eth-trunk 1
interface GigabitEthernet0/0/4
    eth-trunk 1
stp instance 1 root primary
```

图 24-13 链路聚合接口

```
vlan batch 2 to 100
interface Ethernet0/0/1
    port link-type trunk
    port trunk allow-pass vlan 2 to 4094
interface Ethernet0/0/2
    port link-type access
    port default vlan 10
    stp edged-port enable
interface Ethernet0/0/3
    port link-type access
    port default vlan 10
    stp edged-port enable
interface Ethernet0/0/4
    port link-type trunk
    port trunk allow-pass vlan 2 to 4094
```

图 24-14 接口配置

```
interface GigabitEthernet0/0/0
    ip address 3.3.3.2 255.255.255.0
interface GigabitEthernet0/0/1
    ip address 4.4.4.2 255.255.255.0
interface LoopBack0
    ip address 8.8.8.8 255.255.255.0
```

图 24-15 运营商接口地址配置

```
rip 1
Version 2
network 3.0.0.0
network 4.0.0.0
network 8.0.0.0
```

图 24-16　RIP 宣告

```
PC11
基础配置  命令行  组播  UDP发包工具  串口
Link local IPv6 address............: fe80::5689:98ff:fe49:4a9c
IPv6 address.......................: :: / 128
IPv6 gateway.......................: ::
IPv4 address.......................: 192.168.10.252
Subnet mask........................: 255.255.255.0
Gateway............................: 192.168.10.1
Physical address...................: 54-89-98-49-4A-9C
DNS server.........................: 192.168.100.101

PC>ping 192.168.10.254

Ping 192.168.10.254: 32 data bytes, Press Ctrl_C to break
From 192.168.10.254: bytes=32 seq=1 ttl=128 time=31 ms
From 192.168.10.254: bytes=32 seq=2 ttl=128 time=47 ms
From 192.168.10.254: bytes=32 seq=3 ttl=128 time=31 ms
From 192.168.10.254: bytes=32 seq=4 ttl=128 time=31 ms
From 192.168.10.254: bytes=32 seq=5 ttl=128 time=47 ms

--- 192.168.10.254 ping statistics ---
  5 packet(s) transmitted
  5 packet(s) received
  0.00% packet loss
  round-trip min/avg/max = 31/37/47 ms
```

图 24-17　同网段间验证

```
PC1
基础配置  命令行  组播  UDP发包工具  串口
Request timeout!
Request timeout!
Request timeout!
Request timeout!

--- 192.168.100.100 ping statistics ---
  5 packet(s) transmitted
  0 packet(s) received
  100.00% packet loss

PC>ping 192.168.100.100

Ping 192.168.100.100: 32 data bytes, Press Ctrl_C to break
From 192.168.100.100: bytes=32 seq=1 ttl=254 time=109 ms
From 192.168.100.100: bytes=32 seq=2 ttl=254 time=62 ms
From 192.168.100.100: bytes=32 seq=3 ttl=254 time=63 ms
From 192.168.100.100: bytes=32 seq=4 ttl=254 time=109 ms
From 192.168.100.100: bytes=32 seq=5 ttl=254 time=47 ms

--- 192.168.100.100 ping statistics ---
  5 packet(s) transmitted
  5 packet(s) received
  0.00% packet loss
  round-trip min/avg/max = 47/78/109 ms
```

图 24-18　客户端访问服务器

（3）客户端访问公网，如图 24-19 所示。

图 24-19　客户端访问公网

（4）查看核心交换机 VRRP，查看 Switch 上的 VRRP，如图 24-20 所示。

```
<CoreA>display vrrp br
VRID  State     Interface    Type    Virtual IP
------------------------------------------------
10    Master    Vlanif10     Normal  192.168.10.1
11    Master    Vlanif11     Normal  192.168.11.1
12    Master    Vlanif12     Normal  192.168.12.1
13    Master    Vlanif13     Normal  192.168.13.1
100   Master    Vlanif100    Normal  192.168.100.1
------------------------------------------------
Total:5    Master:5    Backup:0    Non-active:0
```

(a) 查看CoreA上的VRRP

```
<CoreB>display vrrp brief
VRID  State     Interface    Type    Virtual IP
------------------------------------------------
10    Backup    Vlanif10     Normal  192.168.10.1
11    Backup    Vlanif11     Normal  192.168.11.1
12    Backup    Vlanif12     Normal  192.168.12.1
13    Backup    Vlanif13     Normal  192.168.13.1
100   Backup    Vlanif100    Normal  192.168.100.1
------------------------------------------------
Total:5    Master:0    Backup:5    Non-active:0
```

(b) 查看CoreB上的VRRP

图 24-20　查看 Switch 上的 VRRP

（5）查看 OSPF 邻居，如图 24-21 所示。

（6）查看 OSPF 路由表，如图 24-22 所示。

（7）查看运营商 RIP 路由表，如图 24-23 所示。

（8）查看核心交换机链路聚合，如图 24-24 所示。

（9）查看防火墙 HRP 状态，如图 24-25 所示。

```
<CoreB>display ospf peer brief

      OSPF Process 100 with Router ID 192.168.2.2
             Peer Statistic Information
----------------------------------------------------------------------------
Area Id          Interface                          Neighbor id       State
0.0.0.0          Vlanif2                            192.168.2.1       Full
0.0.0.0          Vlanif4                            192.168.1.1       Full
0.0.0.0          Vlanif10                           192.168.1.2       Full
0.0.0.0          Vlanif11                           192.168.1.2       Full
0.0.0.0          Vlanif12                           192.168.1.2       Full
0.0.0.0          Vlanif13                           192.168.1.2       Full
0.0.0.0          Vlanif100                          192.168.1.2       Full
----------------------------------------------------------------------------
```

图 24-21　查看 OSPF 邻居

```
<CoreA>display ip routing-table
Route Flags: R - relay, D - download to fib
------------------------------------------------------------------------------
Routing Tables: Public
         Destinations : 25          Routes : 36

Destination/Mask    Proto   Pre  Cost      Flags NextHop         Interface

        0.0.0.0/0   Static  60   0           RD  192.168.1.1     Vlanif1
      127.0.0.0/8   Direct  0    0            D  127.0.0.1       InLoopBack0
      127.0.0.1/32  Direct  0    0            D  127.0.0.1       InLoopBack0
    192.168.1.0/24  Direct  0    0            D  192.168.1.2     Vlanif1
    192.168.1.2/32  Direct  0    0            D  127.0.0.1       Vlanif1
    192.168.2.0/24  OSPF    10   2            D  192.168.13.3    Vlanif13
                    OSPF    10   2            D  192.168.12.3    Vlanif12
                    OSPF    10   2            D  192.168.10.3    Vlanif10
                    OSPF    10   2            D  192.168.100.3   Vlanif100
                    OSPF    10   2            D  192.168.3.1     Vlanif3
                    OSPF    10   2            D  192.168.11.3    Vlanif11
    192.168.3.0/24  Direct  0    0            D  192.168.3.2     Vlanif3
    192.168.3.2/32  Direct  0    0            D  127.0.0.1       Vlanif3
    192.168.4.0/24  OSPF    10   2            D  192.168.13.3    Vlanif13
                    OSPF    10   2            D  192.168.12.3    Vlanif12
                    OSPF    10   2            D  192.168.10.3    Vlanif10
                    OSPF    10   2            D  192.168.100.3   Vlanif100
                    OSPF    10   2            D  192.168.1.1     Vlanif1
                    OSPF    10   2            D  192.168.11.3    Vlanif11
    192.168.5.0/24  OSPF    10   2            D  192.168.3.1     Vlanif3
                    OSPF    10   2            D  192.168.1.1     Vlanif1
   192.168.10.0/24  Direct  0    0            D  192.168.10.2    Vlanif10
   192.168.10.1/32  Direct  0    0            D  127.0.0.1       Vlanif10
   192.168.10.2/32  Direct  0    0            D  127.0.0.1       Vlanif10
   192.168.11.0/24  Direct  0    0            D  192.168.11.2    Vlanif11
   192.168.11.1/32  Direct  0    0            D  127.0.0.1       Vlanif11
```

图 24-22　查看 OSPF 路由表

（10）查看防火墙攻击防护，如图 24-26 所示。

五、实验总结

本实验能够让学生进一步认识到防火墙设备在网络中的作用，学会对防火墙进行基本配置、基本安全配置、域间策略配置、攻击防护配置等。

```
<ISPA>display ip routing-table pr
<ISPA>display ip routing-table protocol r
<ISPA>display ip routing-table protocol rip
Route Flags: R - relay, D - download to fib
------------------------------------------------------------
Public routing table : RIP
        Destinations : 3        Routes : 3

RIP routing table status : <Active>
        Destinations : 3        Routes : 3

Destination/Mask    Proto   Pre   Cost      Flags  NextHop    Interface

     2.2.2.0/24     RIP     100   2           D    3.3.3.2    GigabitEtherne
0/0/1
     4.4.4.0/24     RIP     100   1           D    3.3.3.2    GigabitEtherne
0/0/1
     8.8.8.0/24     RIP     100   1           D    3.3.3.2    GigabitEtherne
0/0/1

RIP routing table status : <Inactive>
        Destinations : 0        Routes : 0
```

图 24-23　查看运营商 RIP 路由表

```
<CoreA>display eth-trunk 1
Eth-Trunk1's state information is:
WorkingMode: NORMAL           Hash arithmetic: According to SIP-XOR-DIP
Least Active-linknumber: 1    Max Bandwidth-affected-linknumber: 8
Operate status: up            Number Of Up Port In Trunk: 2
--------------------------------------------------------------
PortName                      Status         Weight
GigabitEthernet0/0/3          Up             1
GigabitEthernet0/0/4          Up             1
```

图 24-24　查看链路聚合

```
HRP_M<FWA>display hrp state
2021-04-08 08:13:06.830
 Role: active, peer: active
 Running priority: 45000, peer: 45000
 Backup channel usage: 0.00%
 Stable time: 0 days, 0 hours, 22 minutes
 Last state change information: 2021-04-08 7:50:17 HRP link changes to up.
```

图 24-25　防火墙 HRP 状态

六、实验思考题

(1) 什么是防火墙，原理是什么？

(2) 防火墙的类型分为几种？

(3) 防火墙的部署方式是什么，有哪几种？

(4) 防火墙分为哪几个区域，各区域分别是做什么的？

```
HRP_M<FWA>display firewall defend flag ?
  <cr>
HRP_M<FWA>display firewall defend flag
2021-04-08 08:13:57.000

    port-scan                  : disable
    ip-sweep                   : disable
    teardrop                   : disable
    time-stamp                 : enable
    route-record               : enable
    source-route               : enable
    ip-fragment                : enable
    tcp-flag                   : disable
    winnuke                    : disable
    fraggle                    : disable
    tracert                    : enable
    icmp-unreachable           : disable
    icmp-unreachable(DF)       : disable
    icmp-redirect              : disable
    large-icmp                 : disable
    ping-of-death              : enable
    smurf                      : enable
    land                       : enable
    ip-spoofing                : disable
    arp-spoofing               : disable

HRP_M<FWA>
Apr  8 2021 08:14:02 FWA DS/4/DATASYNC_CFGCHANGE:OID 1.3.6.1.4.1.2011.5.25.191
```

图 24-26　查看防火墙攻击防护

实验二十五 典型企业网络的设计与实现

互联网的普及,对于各行各业的发展带来了特别重要的影响。很多公司内部已经实现了信息化,而一些大中型企业早已经构建出局域网和广域网基本接入的网络体系,这样做不仅可以加强企业网内部各部门间的管理工作,对于各部门间安全也起到了非常好的保护作用,同时提高了各部门间的工作效率,节省了成本,为企业在各大领域的发展提供了新的机遇。

企业网的建设应该秉承以下几条基本原则。

(1)开放性、高效性。

在大中型企业网络建设时,选择当前最为可靠的网络技术,保证在未来一段时间不被市场淘汰。采用统一的网络协议和接口标准,来实现企业内部网络的互联,可以在企业内部形成局域网。

(2)高性价比。

在大中型企业的发展过程中,企业网络建设是一种必然的发展需求,需要以低成本、高回报的原则来完成建设。

(3)建设科学合理的企业网络管理机制。

可以帮助企业完善管理制度、管理平台,编撰适合企业且符合标准的规章制度手册。

大中型企业相当重视网络拓扑的结构设计,希望建立一个范围大、功能强、灵活性高、稳定性强的网络,因此大都采用分层网络设计的原则。分层网络设计层次清楚、结构简单、可扩展性好。随着企业规模的扩大,相关部门增多,网络的扩展也比较容易实现,新增网络的分支不会影响核心骨干网的正常运行。一般来说,大中型企业的网络建设,网络层次主要分为 3 个逻辑服务单元,即核心骨干网、汇聚网和接入网。汇聚网和接入网相当于核心骨干网之外的延伸,它们在结构上是独立于核心骨干网的,因此,无论增加多少个接入端口使用网络,都不会影响核心骨干网。所以,对于大中型企业来说,都采用这种分层网络设计结构。

一、实验目的

(1)掌握 OSPF 协议的常用配置方法。

(2)掌握 AC6005 的常用配置方法。

(3) 掌握路由器 NAT 的常用配置方法。
(4) 掌握 DHCP 的常用配置方法。

二、实验环境

eNSP V100R003C00SPC100 模拟器(路由器、交换机、直通线、计算机)。

三、实验内容及步骤

(1) 搭建实验拓扑。
(2) 配置接口 IP 信息。
(3) 配置 OSPF 协议。
(4) 配置 NAT 转换。
(5) 配置 AC6005。
(6) 查看各路由器路由表。
(7) 网络测试。
(8) 查看 AP 上线情况。

四、实验过程

1. 实验拓扑

根据实验内容,设计网络拓扑,如图 25-1 所示。

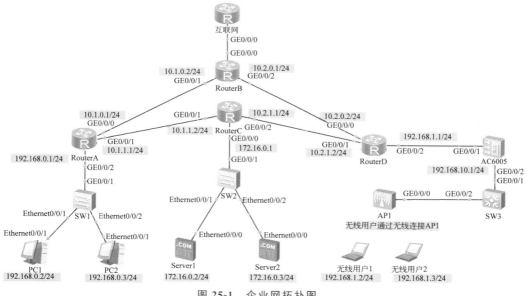

图 25-1 企业网拓扑图

2. 地址规划

参照拓扑图,合理规划 IP 地址,如表 25-1 所示。

表 25-1　IP 地址规划表

设　　备	接　　口	IP 地　址
RouterA	GE0/0/1	10.1.1.1/24
RouterA	GE0/0/2	192.168.0.1/24
RouterB	GE0/0/0	DHCP
RouterB	GE0/0/1	10.1.0.2/24
RouterB	GE0/0/2	10.2.0.1/24
RouterC	GE0/0/1	10.1.1.2/24
RouterC	GE0/0/2	10.2.1.1/24
RouterD	GE0/0/0	10.2.0.2/24
RouterD	GE0/0/1	10.2.1.2/24
RouterD	GE0/0/2	192.168.1.1/24
互联网	GE0/0/0	192.168.18.1/24
PC1	Eth0/0/1	192.168.0.2/24
PC2	Eth0/0/1	192.168.0.3/24
Server1	Eth0/0/0	172.16.0.2/24
Server2	Eth0/0/0	172.16.0.3/24
AC6005	VLAN 20	192.168.10.1/24
无线用户 1		192.168.1.2/24
无线用户 2		192.168.1.3/24

3．实验配置

（1）配置 RouterA ID，配置接口地址，开启 OSPF 协议，配置广播地址，保存配置，如图 25-2 所示。

```
[RouterA]router id 1.1.1.1
[RouterA]interface GigabitEthernet0/0/0
[RouterA-GigabitEthernet0/0/0]ip address 10.1.0.1 255.255.255.0
[RouterA]interface GigabitEthernet0/0/1
[RouterA-GigabitEthernet0/0/1]ip address 10.1.1.1 255.255.255.0
[RouterA]interface GigabitEthernet0/0/2
[RouterA-GigabitEthernet0/0/2]ip address 192.168.0.1 255.255.255.0
[RouterA]ospf 1
[RouterA-ospf-1]area 0.0.0.0
[RouterA-ospf-1-area-0.0.0.0]network 10.1.0.0 0.0.0.255
[RouterA-ospf-1-area-0.0.0.0]network 10.1.1.0 0.0.0.255
[RouterA-ospf-1-area-0.0.0.0]network 192.168.0.0 0.0.0.255
<RouterA>save
```

图 25-2　RouterA 配置

（2）配置 RouterB ID，开启全局 DHCP，配置 ACL2999 放行所有，配置接口地址，配置广播地址，默认路由通告，保存配置，如图 25-3 所示。

```
[RouterB]router id 2.2.2.2
[RouterB]dhcp enable
[RouterB]acl number 2999
[RouterB-acl-basic-2999]rule permit
[RouterB]interface GigabitEthernet 0/0/0
[RouterB-GigabitEthernet0/0/0]ip address dhcp-alloc
[RouterB-GigabitEthernet0/0/0]nat outbound 2999
[RouterB]interface GigabitEthernet0/0/1
[RouterB-GigabitEthernet0/0/1]ip address 10.1.0.2 255.255.255.0
[RouterB]interface GigabitEthernet0/0/2
[RouterB-GigabitEthernet0/0/2]ip address 10.2.0.1 255.255.255.0
[RouterB]ip route-static 0.0.0.0 0.0.0.0 GigabitEthernet 0/0/0
[RouterB]ospf 1
[RouterB-ospf-1]area 0.0.0.0
[RouterB-ospf-1-area-0.0.0.0]network 10.1.0.0 0.0.0.255
[RouterB-ospf-1-area-0.0.0.0]network 10.2.0.0 0.0.0.255
[RouterB-ospf-1]default-route-advertise
<RouterB>save
```

图 25-3　RouterB 配置

（3）配置 RouterC ID 接口，配置接口地址，开启 OSPF 协议，配置广播地址，保存配置，如图 25-4 所示。

```
[RouterC]router id 3.3.3.3
[RouterC]interface GigabitEthernet0/0/0
[RouterC-GigabitEthernet0/0/0]ip address 172.16.0.1 255.255.255.0
[RouterC]interface GigabitEthernet0/0/1
[RouterC-GigabitEthernet0/0/1]ip address 10.1.1.2 255.255.255.0
[RouterC]interface GigabitEthernet0/0/2
[RouterC-GigabitEthernet0/0/2]ip address 10.2.1.1 255.255.255.0
[RouterC]ospf 1
[RouterC-ospf-1]area 0.0.0.0
[RouterC-ospf-1-area-0.0.0.0]network 10.1.1.0 0.0.0.255
[RouterC-ospf-1-area-0.0.0.0]network 10.2.1.0 0.0.0.255
<RouterC>save
```

图 25-4　RouterC 配置

（4）配置 RouterD ID 接口，开启全局 DHCP 功能，配置接口地址，启用 OSPF 协议，配置广播地址，保存配置，如图 25-5 所示。

（5）配置互联网，开启全局 DHCP 功能，配置接口地址，启用 DHCP 服务自动获取 DNS 列表，保存配置。

（6）AC6005 的配置：创建 VLAN，开启全局 DHCP 功能、启用接口，接口配置 access 为 VLAN 10，如图 25-6 所示。

（7）在 AC6005 上建立 CAPWAP 隧道，使用的接口为 VLANIF 10，进入无线配置模式配置 AP 为不认证模式，配置 SSID 模版，如图 25-7 所示。

```
[RouterD]dhcp enable
[RouterD]router id 4.4.4.4
[RouterD]interface GigabitEthernet0/0/0
[RouterD-GigabitEthernet0/0/0]ip address 10.2.0.2 255.255.255.0
[RouterD]interface GigabitEthernet0/0/1
[RouterD-GigabitEthernet0/0/1]ip address 10.2.1.2 255.255.255.0
[RouterD]interface GigabitEthernet0/0/2
[RouterD-GigabitEthernet0/0/2]ip address 192.168.1.1 255.255.255.0
[RouterD]ospf 1
[RouterD-ospf-1]area 0.0.0.0
[RouterD-ospf-1-area-0.0.0.0]network 10.2.0.0 0.0.0.255
[RouterD-ospf-1-area-0.0.0.0]network 10.2.1.0 0.0.0.255
[RouterD-ospf-1-area-0.0.0.0]network 192.168.1.0 0.0.0.255
<RouterD>save
```

图 25-5 RouterD 配置

```
[AC6005]vlan 10
[AC6005]dhcp enable
[AC6005]interface Vlanif 10
[AC6005-Vlanif10]ip address 192.168.10.1 255.255.255.0
[AC6005-Vlanif10]dhcp select interface
[AC6005]interface GigabitEthernet 0/0/2
[AC6005-GigabitEthernet0/0/2]port link-type access
[AC6005-GigabitEthernet0/0/2]port default vlan 10
```

图 25-6 AC 基础配置

```
[AC6005]capwap source interface Vlanif 10
[AC6005]wlan
[AC6005-wlan-view]ap auth-mode no-auth
[AC6005-wlan-view]ssid-profile name shiyan
[AC6005-wlan-ssid-prof-shiyan]ssid shiyan
[AC6005-wlan-ssid-prof-shiyan]quit
```

图 25-7 AC 无线配置

(8) 在 AC6005 上配置无线安全模版并设置密码，配置 VAP 模版关联 SSID 和安全模版，如图 25-8 所示。

```
[AC6005-wlan-view]security-profile name shiyan
[AC6005-wlan-sec-prof-shiyan]security wpa-wpa2 psk pass-phrase shiyan1234 aes
[AC6005-wlan-sec-prof-shiyan]quit
[AC6005-wlan-view]vap-profile name shiyan
[AC6005-wlan-vap-prof-shiyan]ssid-profile shiyan
[AC6005-wlan-vap-prof-shiyan]security-profile shiyan
```

图 25-8 AC 密码配置

(9) 配置 AC6005 为隧道转发模式，配置业务 VLAN 为 VLAN 1，配置 AP 组关联 VAP 模版，保存配置，如图 25-9 所示。

```
[AC6005-wlan-vap-prof-shiyan]forward-mode tunnel
[AC6005-wlan-vap-prof-shiyan]service-vlan vlan-id 1
[AC6005-wlan-vap-prof-shiyan]quit
[AC6005-wlan-view]ap-group name shiyan
[AC6005-wlan-ap-group-shiyan]vap-profile shiyan wlan 1 radio all
[AC6005-wlan-ap-group-shiyan]quit
<AC6005>save
```

图 25-9　AC 隧道配置

4. 结果验证

（1）查看 RouterA 路由表，如图 25-10 所示。

```
<RouterA>disp ip routing-table
Route Flags: R - relay, D - download to fib
------------------------------------------------------------

Routing Tables: Public
        Destinations : 18      Routes : 19

Destination/Mask      Proto    Pre   Cost    Flags NextHop      Interface

0.0.0.0/0             O_ASE    150   1       D     10.1.0.2     GigabitEthernet 0/0/0
10.1.0.0/24           Direct   0     0       D     10.1.0.1     GigabitEthernet 0/0/0
10.1.0.1/32           Direct   0     0       D     127.0.0.1    GigabitEthernet 0/0/0
10.1.0.255/32         Direct   0     0       D     127.0.0.1    GigabitEthernet 0/0/0
10.1.1.0/24           Direct   0     0       D     10.1.1.1     GigabitEthernet 0/0/1
10.1.1.1/32           Direct   0     0       D     127.0.0.1    GigabitEthernet 0/0/1
10.1.1.255/32         Direct   0     0       D     127.0.0.1    GigabitEthernet 0/0/1
10.2.0.0/24           OSPF     10    2       D     10.1.0.2     GigabitEthernet 0/0/0
10.2.1.0/24           OSPF     10    2       D     10.1.1.2     GigabitEthernet 0/0/1
127.0.0.0/8           Direct   0     0       D     127.0.0.1    InLoopBack0
127.0.0.1/32          Direct   0     0       D     127.0.0.1    InLoopBack0
127.255.255.255/32    Direct   0     0       D     127.0.0.1    InLoopBack0
172.16.0.0/24         OSPF     10    2       D     10.1.1.2     GigabitEthernet 0/0/1
192.168.0.0/24        Direct   0     0       D     192.168.0.1  GigabitEthernet 0/0/2
192.168.0.1/32        Direct   0     0       D     127.0.0.1    GigabitEthernet 0/0/2
192.168.0.255/32      Direct   0     0       D     127.0.0.1    GigabitEthernet 0/0/2
192.168.1.0/24        OSPF     10    3       D     10.1.1.2     GigabitEthernet 0/0/1
                      OSPF     10    3       D     10.1.0.2     GigabitEthernet 0/0/0
255.255.255.255/32    Direct   0     0       D     127.0.0.1    InLoopBack0
```

图 25-10　RouterA 路由表

（2）查看 RouterB 路由表，如图 25-11 所示。

（3）查看 RouterC 路由表，如图 25-12 所示。

（4）查看 RouterD 路由表，如图 25-13 所示。

（5）查看 AC6005 下的 AP 是否正常，如图 25-14 所示。

```
<RouterB>disp ip routing-table
Route Flags: R - relay, D - download to fib
------------------------------------------------------------
Routing Tables: Public
        Destinations : 19    Routes : 20

Destination/Mask      Proto   Pre   Cost    Flags NextHop        Interface
        0.0.0.0/0     Static  60    0       D     192.168.18.254 GigabitEthernet0/0/0
        10.1.0.0/24   Direct  0     0       D     10.1.0.2       GigabitEthernet0/0/1
        10.1.0.2/32   Direct  0     0       D     127.0.0.1      GigabitEthernet0/0/1
        10.1.0.255/32 Direct  0     0       D     127.0.0.1      GigabitEthernet0/0/1
        10.1.1.0/24   OSPF    10    2       D     10.1.0.1       GigabitEthernet0/0/1
        10.2.0.0/24   Direct  0     0       D     10.2.0.1       GigabitEthernet0/0/2
        10.2.0.1/32   Direct  0     0       D     127.0.0.1      GigabitEthernet0/0/2
        10.2.0.255/32 Direct  0     0       D     127.0.0.1      GigabitEthernet0/0/2
        10.2.1.0/24   OSPF    10    2       D     10.2.0.2       GigabitEthernet0/0/2
        127.0.0.0/8   Direct  0     0       D     127.0.0.1      InLoopBack0
        127.0.0.1/32  Direct  0     0       D     127.0.0.1      InLoopBack0
     127.255.255.255/32 Direct 0    0       D     127.0.0.1      InLoopBack0
        172.16.0.0/24 OSPF    10    3       D     10.1.0.1       GigabitEthernet0/0/1
                      OSPF    10    3       D     10.2.0.2       GigabitEthernet0/0/2
        192.168.0.0/24 OSPF   10    2       D     10.1.0.1       GigabitEthernet0/0/1
        192.168.1.0/24 OSPF   10    2       D     10.2.0.2       GigabitEthernet0/0/2
        192.168.18.0/24 Direct 0    0       D     192.168.18.254 GigabitEthernet0/0/0
        192.168.18.254/32 Direct 0  0       D     127.0.0.1      GigabitEthernet0/0/0
        192.168.18.255/32 Direct 0  0       D     127.0.0.1      GigabitEthernet0/0/0
        255.255.255.255/32 Direct 0 0       D     127.0.0.1      InLoopBack0
```

图 25-11　RouterB 路由表

```
<RouterC>disp ip routing-table
Route Flags: R - relay, D - download to fib
------------------------------------------------------------
Routing Tables: Public
        Destinations : 18    Routes : 19

Destination/Mask      Proto   Pre   Cost    Flags NextHop        Interface
        0.0.0.0/0     O_ASE   150   1       D     10.1.1.1       GigabitEthernet0/0/1
                      O_ASE   150   1       D     10.2.1.2       GigabitEthernet0/0/2
        10.1.0.0/24   OSPF    10    2       D     10.1.1.1       GigabitEthernet0/0/1
        10.1.1.0/24   Direct  0     0       D     10.1.1.2       GigabitEthernet0/0/1
        10.1.1.2/32   Direct  0     0       D     127.0.0.1      GigabitEthernet0/0/1
        10.1.1.255/32 Direct  0     0       D     127.0.0.1      GigabitEthernet0/0/1
        10.2.0.0/24   OSPF    10    2       D     10.2.1.2       GigabitEthernet0/0/2
        10.2.1.0/24   Direct  0     0       D     10.2.1.1       GigabitEthernet0/0/2
        10.2.1.1/32   Direct  0     0       D     127.0.0.1      GigabitEthernet0/0/2
        10.2.1.255/32 Direct  0     0       D     127.0.0.1      GigabitEthernet0/0/2
        127.0.0.0/8   Direct  0     0       D     127.0.0.1      InLoopBack0
        127.0.0.1/32  Direct  0     0       D     127.0.0.1      InLoopBack0
     127.255.255.255/32 Direct 0    0       D     127.0.0.1      InLoopBack0
        172.16.0.0/24 Direct  0     0       D     172.16.0.1     GigabitEthernet0/0/0
        172.16.0.1/32 Direct  0     0       D     127.0.0.1      GigabitEthernet0/0/0
        172.16.0.255/32 Direct 0    0       D     127.0.0.1      GigabitEthernet0/0/0
        192.168.0.0/24 OSPF   10    2       D     10.1.1.1       GigabitEthernet0/0/1
        192.168.1.0/24 OSPF   10    2       D     10.2.1.2       GigabitEthernet0/0/2
        255.255.255.255/32 Direct 0 0       D     127.0.0.1      InLoopBack0
```

图 25-12　RouterC 路由表

```
<RouterD>disp ip routing-table
Route Flags: R - relay, D - download to fib
------------------------------------------------------------
Routing Tables: Public
        Destinations : 18      Routes : 19
Destination/Mask    Proto   Pre   Cost    Flags  NextHop         Interface

      0.0.0.0/0    O_ASE    150    1        D    10.2.0.1        GigabitEthernet0/0/0
     10.1.0.0/24   OSPF     10     2        D    10.2.0.1        GigabitEthernet0/0/0
     10.1.1.0/24   OSPF     10     2        D    10.2.1.1        GigabitEthernet0/0/1
     10.2.0.0/24   Direct   0      0        D    10.2.0.2        GigabitEthernet0/0/0
     10.2.0.2/32   Direct   0      0        D    127.0.0.1       GigabitEthernet0/0/0
   10.2.0.255/32   Direct   0      0        D    127.0.0.1       GigabitEthernet0/0/0
     10.2.1.0/24   Direct   0      0        D    10.2.1.2        GigabitEthernet0/0/1
     10.2.1.2/32   Direct   0      0        D    127.0.0.1       GigabitEthernet0/0/1
   10.2.1.255/32   Direct   0      0        D    127.0.0.1       GigabitEthernet0/0/1
     127.0.0.0/8   Direct   0      0        D    127.0.0.1       InLoopBack0
     127.0.0.1/32  Direct   0      0        D    127.0.0.1       InLoopBack0
 127.255.255.255/32 Direct  0      0        D    127.0.0.1       InLoopBack0
     172.16.0.0/24 OSPF     10     2        D    10.2.1.1        GigabitEthernet0/0/1
    192.168.0.0/24 OSPF     10     3        D    10.2.1.1        GigabitEthernet0/0/1
                   OSPF     10     3        D    10.2.0.1        GigabitEthernet0/0/0
    192.168.1.0/24 Direct   0      0        D    192.168.1.1     GigabitEthernet0/0/2
    192.168.1.1/32 Direct   0      0        D    127.0.0.1       GigabitEthernet0/0/2
  192.168.1.255/32 Direct   0      0        D    127.0.0.1       GigabitEthernet0/0/2
  255.255.255.255/32 Direct 0      0        D    127.0.0.1       InLoopBack0
```

图 25-13 RouterD 路由表

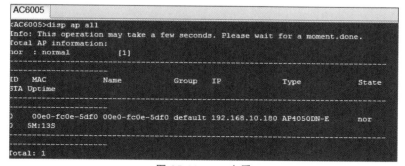

图 25-14 AP 查看

五、实验总结

本实验可以让学生进一步了解 OSPF 技术、路由器 NAT 技术、无线 AC 技术在网络中的应用,学会搭建企业组网环境,保证企业业务正常运行。

六、实验思考题

(1) 企业网络拓扑设计应该遵循什么原则,该如何设计?

(2) 网络出现问题无法上网,应该如何排除故障?

参考文献

[1] 李联宁. 网络工程[M]. 2版. 北京：清华大学出版社,2017.
[2] 李联宁. 网络工程[M]. 3版. 北京：清华大学出版社,2021.
[3] LAMMLE T. CCNA学习指南[M]. 黄国忠,徐宏,译. 7版. 北京：人民邮电出版社,2012.
[4] 杨功元. Packet Tracer使用指南及实验实训教程[M]. 北京：电子工业出版社,2017.
[5] 沈鑫剡. 计算机网络工程[M]. 北京：清华大学出版社,2013.
[6] 吴功宜,吴英. 计算机网络[M]. 4版. 北京：清华大学出版社,2017.

图书资源支持

感谢您一直以来对清华版图书的支持和爱护。为了配合本书的使用,本书提供配套的资源,有需求的读者请扫描下方的"书圈"微信公众号二维码,在图书专区下载,也可以拨打电话或发送电子邮件咨询。

如果您在使用本书的过程中遇到了什么问题,或者有相关图书出版计划,也请您发邮件告诉我们,以便我们更好地为您服务。

我们的联系方式:

地　　址:北京市海淀区双清路学研大厦 A 座 714

邮　　编:100084

电　　话:010-83470236　010-83470237

客服邮箱:2301891038@qq.com

QQ:2301891038(请写明您的单位和姓名)

资源下载:关注公众号"书圈"下载配套资源。

书圈

清华计算机学堂

观看课程直播